Groundwater Simulation and Management Models for the Upper Klamath Basin, Oregon and California

By Marshall W. Gannett, Brian J. Wagner, and Kenneth E. Lite, Jr.

Prepared in cooperation with the Bureau of Reclamation and the Oregon Water Resources Department

Scientific Investigations Report 2012–5062

U.S. Department of the Interior
U.S. Geological Survey

U.S. Department of the Interior
KEN SALAZAR, Secretary

U.S. Geological Survey
Marcia K. McNutt, Director

U.S. Geological Survey, Reston, Virginia: 2012

For more information on the USGS—the Federal source for science about the Earth, its natural and living resources, natural hazards, and the environment, visit http://www.usgs.gov or call 1–888–ASK–USGS.

For an overview of USGS information products, including maps, imagery, and publications, visit http://www.usgs.gov/pubprod

To order this and other USGS information products, visit http://store.usgs.gov

Suggested citation:
Gannett, M.W., Wagner, B.J., and Lite, K.E., Jr., 2012, Groundwater simulation and management models for the upper Klamath Basin, Oregon and California: U.S. Geological Survey Scientific Investigations Report 2012–5062, 92 p.

Contents

Contents—Continued

Figures

Figures—Continued

Figures—Continued

Figures—Continued

Tables

Conversion Factors, Datums, and Abbreviations and Acronyms

Conversion Factors

Multiply	By	To obtain
Length		
inch (in.)	2.54	centimeter (cm)
inch (in.)	25.4	millimeter (mm)
foot (ft)	0.3048	meter (m)
mile (mi)	1.609	kilometer (km)
Area		
acre	4,047	square meter (m^2)
acre	0.4047	hectare (ha)
acre	0.4047	square hectometer (hm^2)
acre	0.004047	square kilometer (km^2)
square foot (ft^2)	929.0	square centimeter (cm^2)
square foot (ft^2)	0.09290	square meter (m^2)
square mile (mi^2)	259.0	hectare (ha)
square mile (mi^2)	2.590	square kilometer (km^2)
Volume		
gallon (gal)	3.785	liter (L)
gallon (gal)	0.003785	cubic meter (m^3)
gallon (gal)	3.785	cubic decimeter (dm^3)
cubic foot (ft^3)	0.02832	cubic meter (m^3)
acre-foot (acre-ft)	1,233	cubic meter (m^3)
acre-foot (acre-ft)	0.001233	cubic hectometer (hm^3)
Flow rate		
gallon per minute (gal/min)	0.0631	Liter per second (L/s)
acre-foot per year (acre-ft/yr)	1,233	cubic meter per year (m^3/yr)
acre-foot per year (acre-ft/yr)	0.001233	cubic hectometer per year (hm^3/yr)
foot per second (ft/s)	0.3048	meter per second (m/s)
foot per year (ft/yr)	0.3048	Meter per year (m/yr)
cubic foot per second (ft^3/s)	0.02832	cubic meter per second (m^3/s)
inch per year (in/yr)	25.4	millimeter per year (mm/yr)
Specific capacity		
gallon per minute per foot [(gal/min)/ft)]	0.2070	liter per second per meter [(L/s)/m]
Hydraulic conductivity		
foot per day (ft/d)	0.3048	meter per day (m/d)
Transmissivity*		
foot squared per day (ft^2/d)	0.09290	meter squared per day (m^2/d)
Leakance		
foot per day per foot [(ft/d)/ft]	1	meter per day per meter

Conversion Factors, Datums, and Abbreviations and Acronyms—Continued

Temperature in degrees Fahrenheit (°F) may be converted to degrees Celsius (°C) as follows:

$$°C=(°F-32)/1.8$$

*Transmissivity: The standard unit for transmissivity is cubic foot per day per square foot times foot of aquifer thickness [(ft³/d)/ft²]ft. In this report, the mathematically reduced form, foot squared per day (ft²/d), is used for convenience.

Datums

Vertical coordinate information is referenced to the National Geodetic Vertical Datum of 1929 (NGVD 29).

Horizontal coordinate information is referenced to the North American Datum of 1927 (NAD 27).

Elevation, as used in this report, refers to distance above the vertical datum.

Abbreviations and Acronyms

AET	actual evapotranspiration
CDWR	California Department of Water Resources
DRN	MODFLOW drain package
ET	evapotranspiration
EVT	MODFLOW evapotranspiration package
GHB	MODFLOW general head boundaries package
GPS	global positioning system
HRU	hydrologic response unit
KBRA	Klamath Basin Restoration Agreement
MODFLOW	MODular three-dimensional finite-difference groundwater FLOW model
OWRD	Oregon Water Resources Department
PET	potential evapotranspiration rate
PRMS	precipitation-runoff modeling system
SBOT	streambed bottom elevation parameter
TID	Tulelake Irrigation District
USGS	U.S. Geological Survey

Groundwater Simulation and Management Models for the Upper Klamath Basin, Oregon and California

By Marshall W. Gannett, Brian J. Wagner, and Kenneth E. Lite, Jr.

Abstract

The upper Klamath Basin encompasses about 8,000 square miles, extending from the Cascade Range east to the Basin and Range geologic province in south-central Oregon and northern California. The geography of the basin is dominated by forested volcanic uplands separated by broad interior basins. Most of the interior basins once held broad shallow lakes and extensive wetlands, but most of these areas have been drained or otherwise modified and are now cultivated. Major parts of the interior basins are managed as wildlife refuges, primarily for migratory waterfowl. The permeable volcanic bedrock of the upper Klamath Basin hosts a substantial regional groundwater system that provides much of the flow to major streams and lakes that, in turn, provide water for wildlife habitat and are the principal source of irrigation water for the basin's agricultural economy.

Increased allocation of surface water for endangered species in the past decade has resulted in increased groundwater pumping and growing interest in the use of groundwater for irrigation. The potential effects of increased groundwater pumping on groundwater levels and discharge to springs and streams has caused concern among groundwater users, wildlife and Tribal interests, and State and Federal resource managers. To provide information on the potential impacts of increased groundwater development and to aid in the development of a groundwater management strategy, the U.S. Geological Survey, in collaboration with the Oregon Water Resources Department and the Bureau of Reclamation, has developed a groundwater model that can simulate the response of the hydrologic system to these new stresses.

The groundwater model was developed using the U.S. Geological Survey MODFLOW finite-difference modeling code and calibrated using inverse methods to transient conditions from 1989 through 2004 with quarterly stress periods. Groundwater recharge and agricultural and municipal pumping are specified for each stress period. All major streams and most major tributaries for which a substantial part of the flow comes from groundwater discharge are included in the model. Groundwater discharge to agricultural drains, evapotranspiration from aquifers in areas of shallow groundwater, and groundwater flow to and from adjacent basins also are simulated in key areas. The model has the capability to calculate the effects of pumping and other external stresses on groundwater levels, discharge to streams, and other boundary fluxes, such as discharge to drains.

Historical data indicate that the groundwater system in the upper Klamath Basin fluctuates in response to decadal climate cycles, with groundwater levels and spring flows rising and declining in response to wet and dry periods. Data also show that groundwater levels fluctuate seasonally and interannually in response to groundwater pumping. The most prominent response is to the marked increase in groundwater pumping starting in 2001. The calibrated model is able to simulate observed decadal-scale climate-driven fluctuations in the groundwater system as well as observed shorter-term pumping-related fluctuations.

Example model simulations show that the timing and location of the effects of groundwater pumping vary markedly depending on the pumping location. Pumping from wells close (within a few miles) to groundwater discharge features, such as springs, drains, and certain streams, can affect those features within weeks or months of the onset of pumping, and the impacts can be essentially fully manifested in several years. Simulations indicate that seasonal variations in pumping rates are buffered by the groundwater system, and peak impacts are closer to mean annual pumping rates than to instantaneous rates. Thus, pumping effects are, to a large degree, spread out over the entire year. When pumping locations are distant (more than several miles) from discharge features, the effects take many years or decades to fully impact those features, and much of the pumped water comes from groundwater storage over a broad geographic area even after two decades. Moreover, because the effects are spread out over a broad area, the impacts to individual features are much smaller than in the case of nearby pumping. Simulations show that the discharge features most affected by pumping in the area of the Bureau of Reclamation's Klamath Irrigation Project are agricultural drains, and impacts to other surface-water features are small in comparison.

A groundwater management model was developed that uses techniques of constrained optimization along with the groundwater flow model to identify the optimal strategy to meet water user needs while not violating defined constraints on impacts to groundwater levels and streamflows. The coupled groundwater simulation-optimization models were formulated to help identify strategies to meet water demand in the upper Klamath Basin. The models maximize groundwater pumping while simultaneously keeping the detrimental impacts of pumping on groundwater levels and groundwater discharge within prescribed limits. Total groundwater withdrawals were calculated under alternative constraints for drawdown, reductions in groundwater discharge to surface water, and water demand to understand the potential benefits and limitations for groundwater development in the upper Klamath Basin.

The simulation-optimization model for the upper Klamath Basin provides an improved understanding of how the groundwater and surface-water system responds to sustained groundwater pumping within the Bureau of Reclamation's Klamath Project. Optimization model results demonstrate that a certain amount of supplemental groundwater pumping can occur without exceeding defined limits on drawdown and stream capture. The results of the different applications of the model demonstrate the importance of identifying constraint limits in order to better define the amount and distribution of groundwater withdrawal that is sustainable.

Background

The upper Klamath Basin spans the Oregon-California border from the flank of the Cascade Range eastward to the high desert (fig. 1). Although much of the basin is high desert, the area receives considerable runoff from the Cascade Range, which forms the western margin, and from the volcanic uplands along the eastern margin. As a result, the area has numerous perennial streams, large shallow lakes, and extensive wetlands. Water in the basin supports irrigation, extensive waterfowl refuges, and aquatic wildlife in lakes and streams throughout the basin.

The agricultural economy of the upper Klamath Basin relies on irrigation water. Just over 500,000 acres are irrigated in the upper Klamath Basin, about 190,000 acres of which are within the Klamath Project developed and operated by the Bureau of Reclamation (Reclamation) (Burt and Freeman, 2003; Natural Resources Conservation Service, 2004). The principal source of water for the Bureau of Reclamation Klamath Project is Upper Klamath Lake. In recent years, Endangered Species Act biological opinions have required Reclamation to maintain certain lake levels in Upper Klamath Lake to protect habitat for endangered fish (specifically the Lost River sucker [*Deltistes luxatus*] and shortnose sucker [*Chasmistes brevirostris*]) and at the same time maintain specified flows in the Klamath River downstream of the lake to provide habitat and suitable conditions for salmon federally listed as threatened. This shift in water management has resulted in increased demands for water. Owing to the limitations of other options, the increased demand has resulted in increased use of groundwater in the basin. The problems associated with increased demands are, of course, exacerbated by drought.

The upper Klamath Basin has a substantial regional groundwater system, and groundwater has been used for irrigation for many decades in certain areas. The changes in water management described above coupled with a series of drier than average years resulted in an approximately 50-percent increase in groundwater pumping between 2000 and 2004 (Gannett and others, 2007). Most of this increase is focused in the area of the Klamath Project. Increased pumping has caused local water-level declines that have been problematic for some groundwater users and have generated concern among resource management agencies and the community. In addition to the measured effects, basic physics requires that the volume of groundwater pumped and used consumptively must ultimately be offset by changes in flow to or from other boundaries including streams (Theis, 1940).

The effects of large-scale groundwater pumping can spread beyond the pumping centers to other parts of the regional groundwater system. Prior to this study, the groundwater hydrology had been studied only in separate parts of the basin, with many areas left undescribed. Therefore, there was no basic framework with which to understand the potential regional effects of groundwater development in the basin and the broad ramifications of water-management decisions. In 1999, the U.S. Geological Survey (USGS) and the Oregon Water Resources Department (OWRD), with assistance from Reclamation, began a cooperative study to quantitatively characterize the regional groundwater system in the upper Klamath Basin and develop a computer model to simulate regional groundwater flow that can be used to help understand the resource and to test management scenarios. This report summarizes the development of the regional groundwater flow model and provides example applications.

Study Objectives

The principal objective of the work described in this report was to provide a numerical model that can be used for the quantitative evaluation of the regional groundwater system in the upper Klamath Basin. The ability to provide quantitative insight into the effects of groundwater pumping on hydraulic heads and discharge to streams, and the response of the groundwater system to decadal climate fluctuations and long-term climate changes was of particular interest. An additional objective was to couple the regional groundwater flow model with a groundwater management model using optimization techniques to identify ways in which groundwater management objectives can be achieved while maintaining hydraulic heads and groundwater discharge rates at needed levels.

Purpose and Scope

The purpose of this report is to describe the development of the upper Klamath Basin regional groundwater flow model and groundwater management models, and to provide example applications. It is intended to help resource managers and other interested parties understand the basic attributes of the models, how they relate to the groundwater hydrology, how they incorporate aspects of groundwater use and water management, and to provide a basic understanding of the theory and application of coupled simulation and groundwater management models.

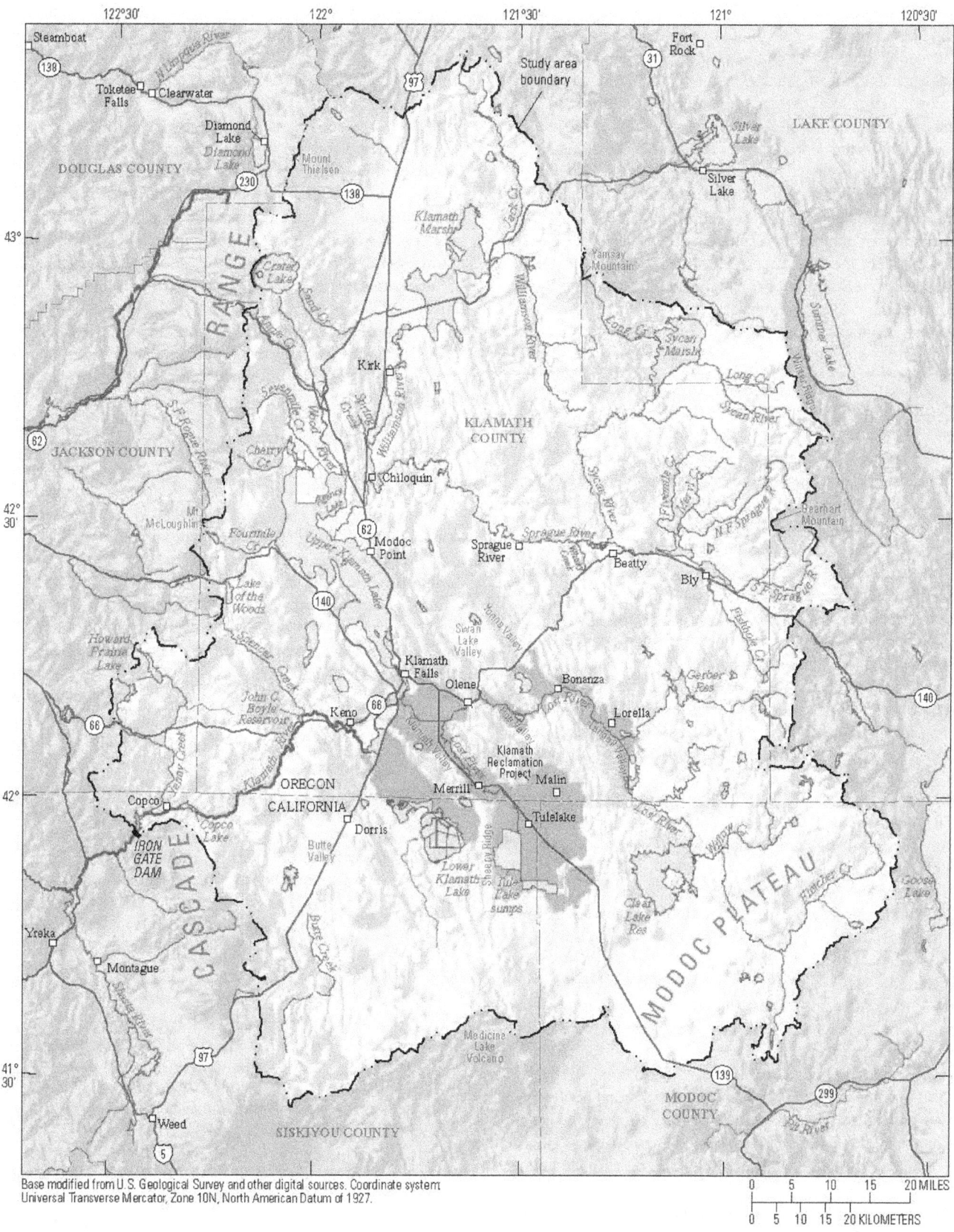

Figure 1. Location of the upper Klamath Basin, Oregon and California, and locations of major geographic features. Bureau of Reclamation Klamath Project shown in orange.

This report briefly describes basic modeling concepts and governing equations, but for a more thorough discussion the reader should refer to a basic groundwater modeling text such as Anderson and Woessner (1992). Much of the report is devoted to the ways in which the basic aspects of the regional groundwater system described by Gannett and others (2007) are represented in the model. The report includes a description of the numerical flow model, its spatial and temporal discretization, the representation of the geologic framework, data and methods used for model calibration, and evaluation of calibration results. The report also includes a discussion of the basic theory of groundwater management models and techniques of constrained optimization in the context of groundwater management in the upper Klamath Basin. Example simulations are included to demonstrate the capabilities of coupled flow and management models.

Study Area Description

The upper Klamath Basin (fig. 1) comprises the entire drainage basin above Iron Gate Dam, including the internally drained Lost River subbasin and Butte Valley area, and encompasses about 8,000 mi². Study-area and model boundaries were defined to correspond to hydrologic boundaries across which groundwater flow can be estimated or assumed to be negligible. The southwestern boundary near Iron Gate Dam was selected because it corresponds with the transition from a geologic terrane dominated by permeable volcanic rock to a terrane dominated by older rock with much lower permeability. There is no significant regional groundwater flow across this geologic boundary.

The boundary between the regional flow systems in the upper Klamath Basin and the Deschutes and Fort Rock Basins to the north (not shown in fig. 1) is defined by a surface-water divide that roughly corresponds to the groundwater divide. This boundary is likely permeable. The boundary between the groundwater system of the upper Klamath Basin and that of the Pit River basin to the south (not shown in fig. 1) also is defined by a surface-water divide in most places. The southern surface-water divide does not correspond to a groundwater divide in all places, as hydraulic head data indicate that there is southward flow of groundwater from the upper Klamath Basin south of the Tule Lake subbasin toward the Pit River basin (Gannett and others, 2007). The eastern study-area boundary corresponds to a surface-water divide and is characterized in many places by a transition to older low-permeability geologic strata.

The upper Klamath Basin occupies a broad, faulted, volcanic plateau that spans the boundary between the Cascade Range and the Basin and Range geologic provinces. The basin is bounded by the volcanic arc of the Cascade Range on the west, the Deschutes River basin to the north, internally drained basins to the east, and the Pit River basin to the south. The elevation of the Cascade Range along the western margin ranges from 5,000 to 7,000 ft with major peaks, such as Mount McLoughlin and Mount Thielsen, exceeding 9,000 ft. The interior parts of the basin are dominated by northwest-trending fault-bounded basins, typically several miles wide, with intervening uplands. Basin floors range in elevation from roughly 4,000 to 4,500 ft, and adjoining fault-block upland elevations range from 4,500 to more than 5,000 ft. The northern and eastern parts of the upper Klamath Basin consist of a volcanic upland with numerous eruptive centers, including Yamsay and Gearhart Mountains, both of which exceed elevations of 8,000 ft. The southeastern margin of the upper Klamath Basin consists of a broad, rugged, volcanic upland known as the Modoc Plateau, where most of the land-surface elevations range from 4,500 to 5,000 ft. The southern margin of the basin is marked by the broad shield of Medicine Lake Volcano, which reaches an elevation of 7,913 ft.

The upper Klamath Basin is semiarid because the Cascade Range intercepts much of the moisture from the predominantly eastward moving Pacific weather systems. Mean annual precipitation (1961–90) ranges from 65.4 in. at Crater Lake National Park in the Cascade Range to 11.1 in. at Tulelake, California (Western Regional Climate Center, 2006) (fig. 2). Most precipitation occurs in the fall and winter. November through March precipitation accounts for 71 percent of the total at Crater Lake and 64 percent of the total at Klamath Falls. Most precipitation falls as snow at high elevations. The interior parts of the basin are very dry during the spring and summer; mean monthly precipitation at Klamath Falls is less than 1 in. from April through October. Winters generally are cold, with January mean-minimum and mean-maximum temperatures of 20.3 and 38.8 °F, respectively, at Klamath Falls and 17.5 °F and 34.5 °F, respectively, at Crater Lake. Summers, in contrast, are warm, with July mean minimum and maximum temperatures of 50.8 °F and 84.6 °F, respectively, at Klamath Falls and 39.8 °F and 68.0 °F, respectively, at Crater Lake.

Principal streams in the upper Klamath Basin include the Williamson River, which drains the northern and eastern parts of the basin; the Sprague River (a tributary to the Williamson), which drains part of the eastern side of the basin; the Lost River, which drains the southeastern part of the basin; and the Klamath River (fig. 1). The Lost River subbasin is actually a closed stream basin. Prior to development, the Lost River flowed to internally drained Tule Lake, although it occasionally received flow from the Klamath River during floods. The Lost River is now diverted just downstream of Olene into a channel across a low divide to the Klamath River. Generally, little water from the Lost River drainage upstream of the diversion channel now flows to the Tule Lake subbasin. Tule Lake is now largely drained except for two connected areas known as the Tule Lake sumps (fig. 1). The largest lake in the basin is Upper Klamath Lake, which has a surface area between 100 and 140 mi² (including non-drained fringe wetlands) depending on stage (Hubbard, 1970; Snyder and Morace, 1997). Principal tributaries to Upper Klamath Lake include the Williamson River, the Wood River (which originates at a series of large springs north of the lake), and several streams emanating from the Cascade Range.

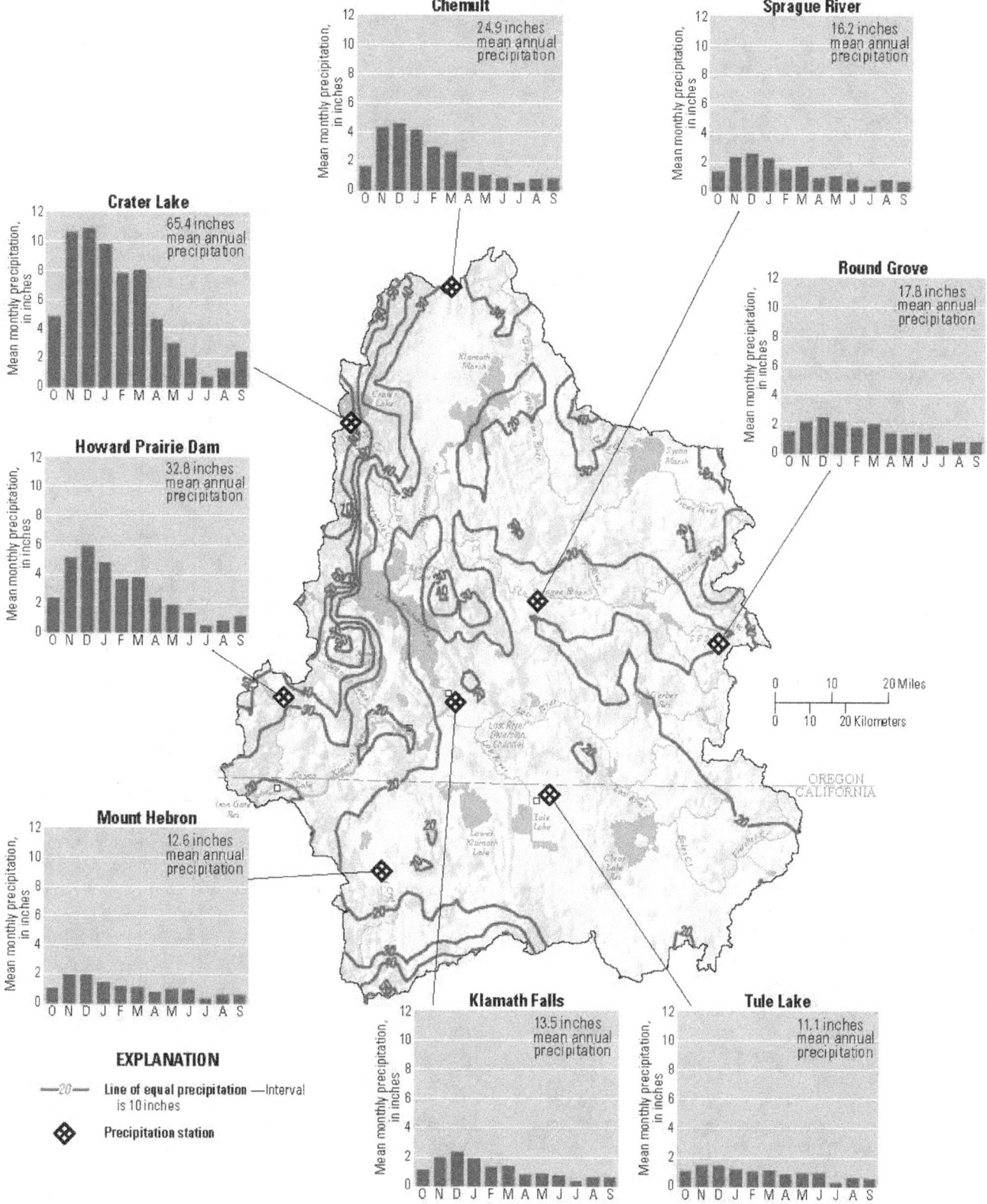

Figure 2. Mean annual precipitation in the upper Klamath Basin, Oregon and California, 1971–2000, and mean monthly precipitation at selected sites, 1961–90. Data from Oregon State University PRISM Group (2006) and Western Regional Climate Center (2006).

The 250-mi-long Klamath River begins at the outlet of Upper Klamath Lake, which is controlled by a dam. For the first mile downstream of the lake, the river is known as the Link River. About 1 mi downstream of the dam, the river flows into a 20-mi-long narrow reservoir behind the dam at Keno known as Lake Ewauna. John C. Boyle Reservoir and its dam are about 10 mi downstream of Keno. Below John C. Boyle Reservoir, the river enters a narrow canyon and flows freely about 20 mi to Copco Lake (a reservoir) and immediately below that, Iron Gate Reservoir. Iron Gate Dam, which impounds Iron Gate Reservoir at about river mile 190, marks the downstream boundary of the upper Klamath Basin. There are no impoundments on the Klamath River downstream of Iron Gate Dam.

The surface hydrology of the upper Klamath Basin has been extensively modified by drainage of lakes and wetlands for agriculture and routing of irrigation water. Prior to development, the Tule Lake and Lower Klamath Lake subbasins contained large lakes fringed by extensive wetlands. Prior to development of the Bureau of Reclamation Klamath Project, the high stage of Tule Lake was about 4,060 ft (La Rue, 1922). At this stage, the lake would cover an area exceeding 150 mi². Historical accounts indicate that at high stage Tule Lake drained into the lava flows along the southern margin. In the early 1900s, the U.S. Reclamation Service (predecessor to the Bureau of Reclamation) experimented with augmenting this subsurface drainage in early attempts to drain the lake. La Rue (1922) reasoned that because the water of Tule Lake was fresh and not saline, the lake "in the past had an outlet." Subsurface drainage also is suggested by the hydraulic head gradient that slopes southward away from the Tule Lake subbasin toward the Pit River Basin. In 1912, a canal and dam were completed that allowed the diversion of water from the Lost River to the Klamath River, cutting off the supply of water to Tule Lake. Most of Tule Lake was drained and is now under cultivation. The only remnants of the lake are the Tule Lake sumps in the southern and western parts of the basin that collect irrigation return flow. Since 1942, water from the sumps has been pumped via a tunnel through Sheepy Ridge into the Lower Klamath Lake subbasin. The Lower Klamath Lake subbasin once held a large lake-marsh complex that covered approximately 88,000 acres, about 58,000 acres of which were marginal wetlands with the remaining 30,000 acres open water (La Rue, 1922). Lower Klamath Lake was connected to the Klamath River through a channel known as the Klamath Strait, and probably through the expansive wetland that separated the lake from the river elsewhere. In the early 1900s, an earth-fill railroad bed was constructed across the northwestern margin of the Lower Klamath Lake subbasin, cutting off flow between the lake and river except at the Klamath Strait. In 1917, the control structure built into this impoundment at the Klamath Strait was closed, cutting off flow to the lake. As a result, Lower Klamath Lake is now largely drained, with much of the former lakebed and fringe wetlands under cultivation. Areas of open water remain in the Lower Klamath Lake Wildlife Refuge in the southern part of the subbasin.

Currently (2011), about 500,000 acres of agricultural land are irrigated in the upper Klamath Basin, roughly 190,000 of which are included in the Bureau of Reclamation Klamath Project (Carlson and Todd, 2003; Natural Resources Conservation Service, 2004). This total does not include wildlife refuge areas within the Project.

The upper Klamath Basin is mostly forested (Loy and others, 2001). Forest trees in upland areas east of the Cascade Range are predominantly ponderosa pine, with areas of true fir and Douglas fir on Yamsay and Gearhart Mountains. Forests in the Cascade Range primarily comprise mountain hemlock and red fir. Lower elevation uplands are dominated by lodgepole pine. Lowland forests consist largely of juniper and sagebrush with some juniper grasslands. Stream valleys and the broad, sediment-filled structural basins generally have extensive marshes, such as Sycan Marsh and Klamath Marsh, except at lower elevations, where the basins have been mostly converted to agricultural land (for example, the Wood River Valley, and the Lower Klamath Lake and Tule Lake subbasins).

Irrigation water comes from various sources in the upper Klamath Basin. Upstream of Upper Klamath Lake, in the Williamson, Sprague, and Wood River drainages, private (non-Project) irrigation water primarily comes from diversion of surface water from the main stem streams or tributaries. A smaller amount of irrigation water is pumped from wells, particularly in the Sprague River Valley and Klamath Marsh areas. In the Langell and Yonna Valleys of the upper Lost River subbasin, irrigation water comes from Clear Lake and Gerber Reservoirs. Irrigators use groundwater and some surface water in Swan Lake Valley. Groundwater is used for irrigation in areas not served by irrigation districts and to supplement surface-water supplies throughout the area.

South of Upper Klamath Lake, most irrigation water comes from the lake, which is the largest single source of irrigation water in the upper Klamath Basin. This area is the main part of the Bureau of Reclamation Klamath Project. Water is stored in and diverted from the lake to irrigate land south of Klamath Falls, including the Klamath Valley, Poe Valley (in the Lost River subbasin), and the Lower Klamath and Tule Lake subbasins. Irrigation return flow (water that originates in Upper Klamath Lake) that ends up in the Tule Lake sumps is pumped through Sheepy Ridge and used for irrigation and refuge purposes in the southern part of the Lower Klamath Lake subbasin. Water diverted from the Klamath River several miles downstream of the lake also is used for irrigation and refuges in the Lower Klamath Lake subbasin. Irrigation and refuge return flow in the Lower Klamath Lake subbasin is routed through a series of pumping stations back to the Klamath River.

A certain amount of groundwater is used for irrigation on land surrounding the Klamath Project upslope of the major canals. Principal areas of groundwater use surrounding the Project area include the southern end of the Klamath Hills, parts of the Klamath Valley, and the northern and eastern margins of the Tule Lake subbasin (fig. 1). Some groundwater traditionally has been used for supplemental irrigation in the Project area. Increased water demand due to drought and requirements for a 100,000 acre-ft pilot water bank to be administered by Reclamation as required by the National Oceanic and Atmospheric Administration Fisheries 2002 biological opinion (National Marine Fisheries Service, 2002) have resulted in a marked increase in groundwater pumping in and around the Klamath Project since 2001 (Gannett and others, 2007).

Hydrogeologic Framework

The groundwater hydrology of the upper Klamath Basin, including the geology, is discussed in detail by Gannett and others (2007). Much of the discussion in this section is from that report.

Geology

The upper Klamath Basin has been a region of volcanic activity for at least 35 million years (Sherrod and Smith, 2000), resulting in complex assemblages of volcanic vents and lava flows, pyroclastic deposits, and volcanically derived sedimentary deposits (fig. 3). Volcanic and tectonic processes have created many of the present-day landforms in the basin. Glaciation and stream processes have subsequently modified the landscape in many places.

The upper Klamath Basin lies within two major geologic provinces, the Cascade Range and the Basin and Range geologic provinces (Orr and others, 1992). The processes that have operated in these provinces have overlapped and interacted in much of the upper Klamath Basin. The Cascade Range is a north-south trending zone of compositionally diverse volcanic eruptive centers with deposits extending from northern California to southern British Columbia. The Cascade Range is subdivided between an older, highly eroded Western Cascades, and a younger, mostly constructional High Cascades. Prominent among the eruptive centers in the High Cascades of the Klamath Basin are large composite and shield volcanoes such as Mount Mazama (Crater Lake), Mount McLoughlin, and Medicine Lake Volcano. The Cascade Range has been impinged on its eastern side by the adjacent structurally-dominated Basin and Range geologic province. The Basin and Range geologic province is a region of crustal extension characterized by subparallel, fault-bounded, down-dropped basins separated by fault-block ranges. Individual basins and intervening ranges are typically 10–20 mi across.

The Basin and Range geologic province encompasses much of the interior of the Western United States, extending from central Oregon southward through Nevada and western Utah, into the southern parts of California, Arizona, and New Mexico. Although the Basin and Range geologic province is primarily structural, faulting has been accompanied by widespread volcanism.

The oldest rocks in the upper Klamath Basin study area are part of the Western Cascades subprovince and consist primarily of early to middle Tertiary lava flows, andesitic mudflows, tuffaceous sedimentary rocks, and vent deposits (fig. 3). The Western Cascade rocks range in age from 20 to 33 million years and are as much as 20,000 ft thick (Hammond, 1983; Vance, 1984). Rocks of the Western Cascades overlie pre-Tertiary rocks of the Klamath Mountains, just west of the study area. Western Cascade rocks have very low permeability because the tuffaceous materials are mostly devitrified (changed to clays and other minerals), and lava flows are weathered and contain abundant secondary minerals. Because of the low permeability, groundwater does not easily move through the Western Cascades rocks, and the unit acts as a barrier to regional groundwater flow. The Western Cascades constitute part of the western boundary of the regional groundwater system. Western Cascade rocks dip toward the east and underlie the High Cascade deposits, defining the lower boundary of the regional flow system throughout that part of the study area.

The High Cascade subprovince ranges in age from late Miocene to late Pleistocene; however, most rocks are Pliocene to Pleistocene in age (Mertzman, 2000). Deposits within the High Cascade subprovince in the study area mostly form constructional features and consist of volcanic vents and lava flows with relatively minor interbedded volcaniclastic and sedimentary deposits. An area of numerous late Miocene to Pliocene cinder cones extends from southwest of Butte Valley to northwest of Mt. Mazama (Crater Lake). Quaternary volcanic deposits are associated with a few volcanic centers concentrated in two general areas in the upper Klamath Basin: from Lake of the Woods north to Crater Lake and from Mt. Shasta (south of the study area) east to Medicine Lake Volcano. The High Cascades rocks are relatively thin in southern Oregon and northern California. High Cascade rocks unconformably overlie Western Cascade rocks and are very permeable, relative to the older rocks.

Deposits in the Basin and Range geologic province in the study area range in age from middle Miocene to Pleistocene. The oldest rocks are middle to late Miocene in age, ranging from 13 to 8 million years. These rocks are exposed just south of the study area in the Pit River Basin and probably underlie the Pliocene age lavas south of Clear Lake Reservoir. The older rocks in the Pit River Basin and bounding the eastern part of the study area are mostly silicic domes, flows, and pyroclastic deposits, which generally have low permeability (California Department of Water Resources, 1963) and typically are faulted and tilted.

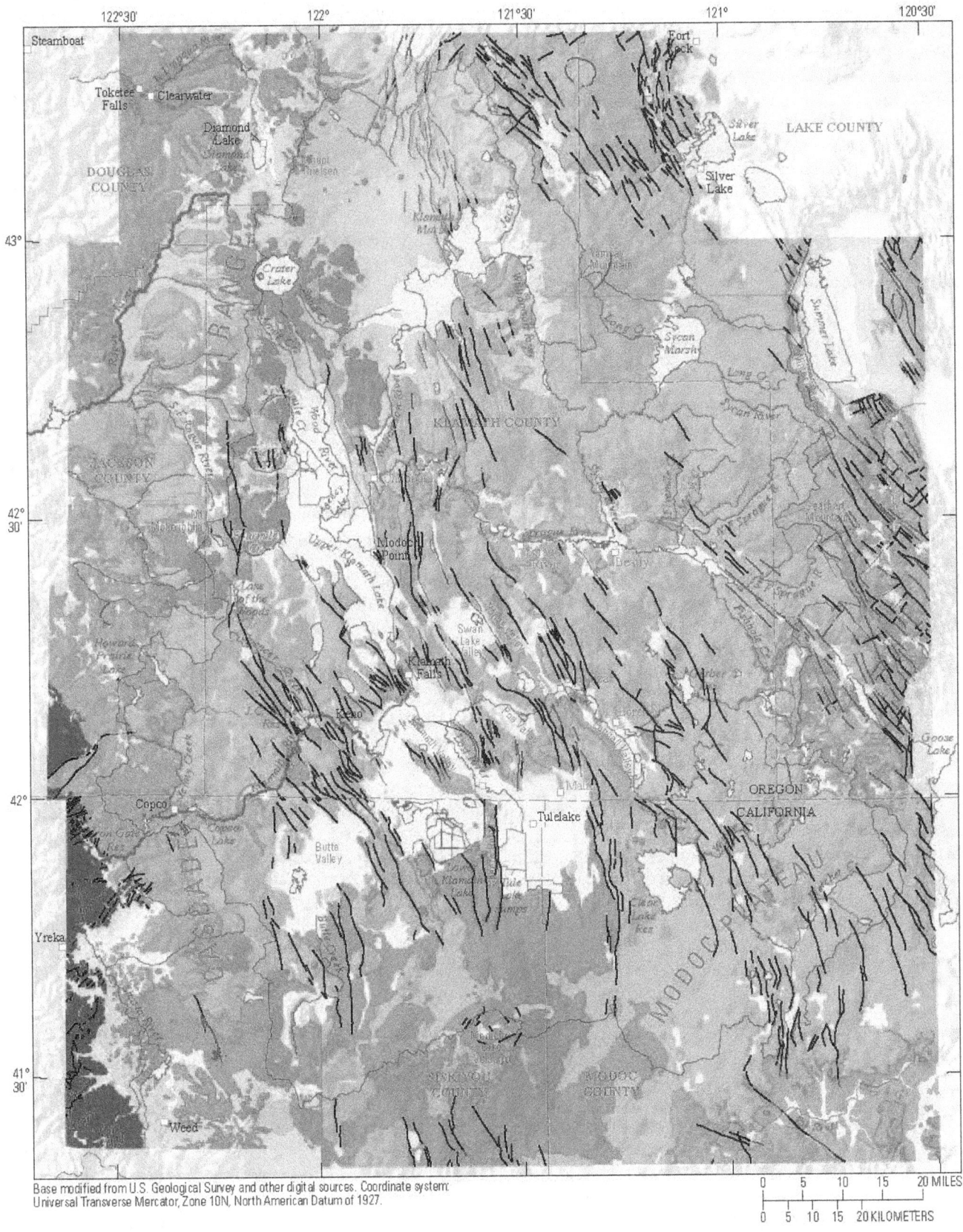

Base modified from U.S. Geological Survey and other digital sources. Coordinate system:
Universal Transverse Mercator, Zone 10N, North American Datum of 1927.

Figure 3.   Generalized geology of the upper Klamath Basin, Oregon and California.

EXPLANATION

Hydrogeologic unit present at land surface

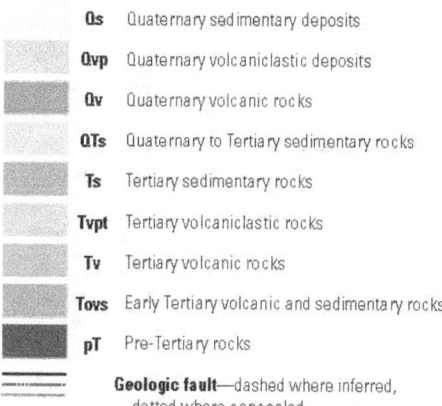

Qs   Quaternary sedimentary deposits

Qvp   Quaternary volcaniclastic deposits

Qv   Quaternary volcanic rocks

QTs   Quaternary to Tertiary sedimentary rocks

Ts   Tertiary sedimentary rocks

Tvpt   Tertiary volcaniclastic rocks

Tv   Tertiary volcanic rocks

Tovs   Early Tertiary volcanic and sedimentary rocks

pT   Pre-Tertiary rocks

Geologic fault—dashed where inferred, dotted where concealed

NOTE: Geology generalized from:
Gay and Aune, 1958;
Walker, 1963;
Smith and others, 1982;
Wagner and Saucedo, 1987;
Sherrod, 1991;
MacLeod and Sherrod, 1992;
Sherrod and Pickthorn, 1992, and
Sherrod and Smith, 2000.

Figure 3.—Continued

Late Miocene to Pliocene volcanic rocks of the Basin and Range geologic province are the major water bearing rocks in the upper Klamath Basin study area. These units consist of volcanic vent deposits and volcanic flow rocks throughout the area east of Upper Klamath Lake and Lower Klamath Lake; the units probably underlie most of the valley- and basin-fill deposits in the study area. Late Miocene to Pliocene rocks also form uplands along the eastern boundary of the study area, and form the plateau that extends from the Langell Valley south to the Pit River. The rocks are predominately basalt and basaltic andesite in composition, but silicic vents and lava flows occur locally, notably in the vicinity of Beatty, Oregon.

Tuff cones and tuff rings are the predominant volcanic vent form in the Sprague River subbasin between Chiloquin and Sprague River, Oregon. Tuff cones and rings form when rising magma comes in contact with water, resulting in explosive fragmentation of the volcanic material and formation of hydrovolcanic deposits. These late Miocene to Pliocene rocks typically exhibit high permeability. Permeability locally may be markedly reduced by secondary mineralization from hydrothermal alteration.

The volcanic rocks of the Basin and Range geologic province are interbedded with, and locally overlain by, late Miocene to Pliocene sedimentary rocks. The sedimentary rocks consist of tuffaceous sandstone, ashy diatomite, mudstone, siltstone, and some conglomerates. These units are exposed both in down-dropped basins and in up-thrown mountain blocks, indicating that the deposits in part represent an earlier generation of sediment-filled basins that have been subsequently faulted and uplifted. These sedimentary deposits are typically poor water producers, and often serve as confining units for underlying volcanic aquifers.

The youngest stratigraphic unit in the upper Klamath Basin consists of late Pliocene to Holocene sedimentary deposits. Those deposits include alluvium along modern flood plains, basin-fill deposits within active grabens, landslide deposits, and glacial drift and outwash. Very thick accumulations of silt, sand, clay, and diatomite underlie the westernmost basins, such as the Upper Klamath Lake, Lower Klamath Lake, Butte Valley, and Tule Lake subbasins. For example, up to 1,740 ft of basin-fill sediment underlies the town of Tulelake, California. Sediment near the base of the deposit at Tulelake has been assigned an age of 3.3 million years on the basis of radiometric ages of interbedded tephra, paleomagnetic data, and estimates of sedimentation rates (Adam and others, 1990). Gravity data suggest that the sediment-fill thickness may exceed 6,000 ft in the Lower Klamath Lake subbasin and may be in the range of 1,300 to 4,000 ft in the Upper Klamath Lake subbasin (Sammel and Peterson, 1976; Veen, 1981; Northwest Geophysical Associates, Inc., 2002; Braunsten, 2009).

Groundwater Hydrology

The upper Klamath Basin has a substantial regional groundwater system. The late Tertiary and Quaternary volcanic rocks that underlie the region are generally permeable, with transmissivity estimates ranging from 1,000 to 100,000 ft^2/d, and compose a system of variously interconnected aquifers. Sedimentary rocks, primarily fine-grained lake sediments and basin-filling deposits, are interbedded with the volcanic rocks. The regional groundwater system is underlain and bounded on the east and west by early Tertiary volcanic and sedimentary rocks that have generally low permeability. Eight regional scale hydrogeologic units are defined in the upper Klamath Basin on the basis of surficial geology and subsurface data (Gannett and others, 2007).

Groundwater flows from recharge areas in the Cascade Range, upland areas in the basin interior, and from eastern margins, toward stream valleys and interior subbasins. Groundwater discharges to streams throughout the basin, and most streams have some component of groundwater discharge (base flow). Some streams, however, are predominantly groundwater fed and have relatively constant flows throughout the year. Large amounts of groundwater discharge to streams in the Wood River subbasin, the lower Williamson River area, and along the margin of the Cascade Range. Much of the inflow to Upper Klamath Lake can be attributed to groundwater discharge to streams and major spring complexes within a dozen or so miles from the lake. This large component of groundwater buffers the lake somewhat from year-to-year variations in annual precipitation, but not from multi-year drought cycles. There are also groundwater discharge areas in the eastern parts of the basin, for example in the upper Williamson and Sprague River subbasins and in the upper Lost River subbasin.

The groundwater system in the upper Klamath Basin responds to external stresses such as climate cycles, pumping, lake-stage variations, and canal operation. This response is manifest as fluctuations in hydraulic head (as represented, for example, by fluctuations in the water-table surface) and variations in groundwater discharge to springs. Basinwide, decadal-scale climate cycles are the largest factor controlling head and discharge fluctuations. Climate-driven water-table fluctuations of more than 12 ft have been observed near the Cascade Range, and decadal-scale fluctuations of 5 ft are common throughout the basin. Groundwater discharge to springs and streams varies throughout the basin by a factor of two or more in response to decadal-scale climate cycles. Climate-driven interannual variations in groundwater discharge total hundreds of cubic feet per second.

The response of the groundwater system to pumping is generally largest in areas of irrigation pumping. Annual drawdown and recovery cycles of 1 to 10 ft are common in pumping areas. Long-term drawdown effects, where the water table has reached or is attempting to reach a new level in equilibrium with the pumping, are apparent in parts of the basin. In general, impacts of pumping on streams and springs are diffuse and difficult to measure. In several instances, however, reductions to spring discharge resulting from nearby pumping are well documented through direct measurement.

Hydraulic Properties of Hydrogeologic Units

Gannett and others (2007) provide a summary of the hydraulic properties (transmissivity and storativity) of regional hydrogeologic units in the upper Klamath Basin based on 32 aquifer tests and specific capacity tests from 288 wells. Transmissivity is the product of the hydraulic conductivity and aquifer thickness. Hydraulic conductivity is the unit volume of water that will move through a unit area of aquifer under a unit hydraulic gradient per unit time, and has dimensions of length per unit time, such as feet per day. Transmissivity then has dimensions of feet squared per day. Storativity is the unit volume of water an aquifer takes into, or releases from, storage per unit area per unit change in head. The volume of water has the units of length cubed (such as ft^3), the area has units of length squared (such as ft^2), and the change in head has units of length (such as ft). Because the volume is divided by the other two quantities, storativity is dimensionless.

Aquifer tests from 26 wells show that the transmissivity of the Tertiary volcanics (predominantly basaltic lavas of Miocene to Pliocene age) varies widely, from 2,700 to 610,000 ft^2/d, with most ranging from 24,000 to 270,000 ft^2/d. The median transmissivity is about 90,000 ft^2/d. Storativity values from aquifer tests in the Tertiary volcanics reported by Gannett and others (2007) range from 0.00001 to 0.15. The 0.15 figure is anomalous and likely due to a partially penetrating observation well and leakage from a confining unit. Most storativity values in the Tertiary volcanics range from 0.00025 to 0.001, and the median is about 0.0005.

Although the number of aquifer tests in Tertiary sediments (or mixtures of the sediments and lavas) is small (6), they provided information on the hydraulic characteristics of the coarse-grained facies of Tertiary sediments. Transmissivity values range from 13,000 to 350,000 ft²/d, with most in the 25,000 to 75,000 ft²/d range. The median value is 54,000 ft²/d. Storativity values range from 0.0005 to 0.015 with most ranging from about 0.0002 to 0.003.

In the early 1980s, the USGS conducted an aquifer test of the geothermal aquifer in Klamath Falls in collaboration with the Lawrence Berkeley Laboratory and the City of Klamath Falls (Benson and others, 1984a, b). The test consisted of four phases: a 1-week pre-test phase during which background water levels were monitored; a 21-day pumping phase during which a geothermal well was pumped at about 720 gal/min and the water discharged to an irrigation canal; a 30-day injection phase during which pumping continued (at about 660–695 gal/min) and the water injected into a second well; and a 1-week recovery phase. Benson and others (1984a) analyzed the data from the test and calculated a permeability-thickness value (analogous to a transmissivity) of about 1.4×10^6 millidarcy-feet. This converts to a transmissivity of about 3,800 ft²/d. Analysis of the test indicated a storativity of about 0.002.

Gannett and others (2007) also summarized transmissivity estimates from specific-capacity tests on 288 water well reports for wells producing from Quaternary sediment, Tertiary sediment, and Tertiary volcanic rock. Wells producing from Quaternary sedimentary deposits and Tertiary sedimentary deposits have similar transmissivity distributions, with the former having slightly larger values. The median transmissivity for both units determined from specific-capacity tests is about 200 ft²/d. The frequency distribution of transmissivities for the Tertiary volcanic deposits is distinct from the other units, with values generally larger by more than an order of magnitude. The median transmissivity of Tertiary volcanic deposits is about 6,300 ft²/d.

The median transmissivity for late Tertiary volcanic deposits determined from specific-capacity tests (6,300 ft²/d) is lower than that calculated from aquifer tests (about 90,000 ft²/d). This is not unexpected for the following reasons. First, transmissivity values determined from single-well tests

can be biased downward by excess drawdown in the pumped well due to well inefficiency (see Driscoll, 1986, p. 244). Aquifer tests with observation wells are not affected by this phenomenon. Second, the large number of specific-capacity tests (173) represents a more or less random sampling of wells (and varying characteristics) in the unit. Aquifer tests, in contrast, are not random but tend to be conducted most commonly on high yielding wells for specific purposes. Regardless, transmissivity values calculated from both aquifer tests and specific-capacity tests are useful for understanding the hydraulic characteristics of hydrogeologic units and the differences between units.

Estimates of the hydraulic properties of the Quaternary volcanic rocks in the Cascade Range are based largely on heat and mass transport models because there are so few wells to provide direct measurements. In simulating groundwater flow and heat transport in the Cascade Range, Ingebritsen and others (1992) estimated the permeability of rocks younger than 2.3 million years to be about 10^{-13} ft², which is equivalent to a hydraulic conductivity of about 0.018 ft/d assuming a water temperature of 41°F. The permeability of rocks with ages between 4 and 8 million years was estimated to be 5.4×10^{-15} ft², which is equivalent to a hydraulic conductivity of about 9.1×10^{-4} ft/d. Higher near-surface permeability, on the order of 0.018 to 1.8 ft/d, was required in their simulation to match groundwater recharge estimates. Higher near-surface permeabilities are also suggested by well-test data. A specific-capacity test of a well near Mount Bachelor yielded a hydraulic-conductivity estimate of 9 ft/d.

Mathematical modeling of groundwater discharge to spring-fed streams in the Cascade Range by Manga (1996, 1997) yielded permeability values for near-surface rocks less than about 2.0 million years old of about 10^{-10} ft², which equates to a hydraulic conductivity of about 18 ft/d, assuming a water temperature of 41°F. This estimate is an order of magnitude larger than the upper value of Ingebritsen and others (1992) for near-surface rocks, where most groundwater flow occurs. The permeability estimates of Manga (1996, 1997) and Ingebritsen and others (1992) are considered to be a reasonable range of values for the younger, near-surface strata in the Cascade Range.

Groundwater Flow Model

Model Description

Governing Equations and Model Code

The movement of groundwater through porous media is described by the following partial differential equation, which is based on Darcy's law and the conservation of mass (McDonald and Harbaugh, 1988):

$$\frac{\partial}{\partial x}\left(K_{xx}\frac{\partial h}{\partial x}\right) + \frac{\partial}{\partial y}\left(K_{yy}\frac{\partial h}{\partial y}\right) + \frac{\partial}{\partial z}\left(K_{zz}\frac{\partial h}{\partial z}\right) - W = S_s\frac{\partial h}{\partial t}, \quad (1)$$

where

$K_{xx}, K_{yy},$

     and K_{zz} are values of hydraulic conductivity in the x, y, and z directions along Cartesian coordinate axes, which are assumed to align with principal directions of hydraulic conductivity (LT^{-1}),

     h is hydraulic head (L),

     W is a volumetric flux per unit volume and represents sinks and/or sources (T^{-1}),

     S_s is the specific storage of the porous material (L^{-1}), and

     t is time (T).

Note that specific storage (S_s) is the storativity (which is dimensionless) divided by the aquifer thickness (which has units of length), resulting in units of one over length (such as ft^{-1}). Derivations of equation (1) can be found in Freeze and Cherry (1979) and Anderson and Woessner (1992). There is no evidence of large-scale horizontal anisotropy in the upper Klamath Basin; therefore, K_{xx} and K_{yy} are considered to be equal at any given location and K_{xx} and K_{yy} are replaced in this discussion by the single term K_h to describe horizontal hydraulic conductivity.

Equation 1 represents the mass balance at a single point in space and time and generally cannot be solved analytically for practical applications involving transient conditions in complex three-dimensional systems. In practice, numerical methods are employed in which the partial differential equation 1 is approximated at a set of spatially discrete points in a process known as discretization. Equation 1 is then replaced by a set of simultaneous algebraic equations that describe the distribution of hydraulic head at each point, and flow through the system in response to this head distribution. These simultaneous equations are set up in matrix form and then solved. A variety of techniques are available to solve the set of simultaneous equations such as the preconditioned conjugate-gradient method of Hill (1990) and the algebraic multigrid solver of Mehl and Hill (2001).

The upper Klamath Basin regional groundwater model was developed using MODFLOW-2000 (Harbaugh and others, 2000; Hill and others, 2000). MODFLOW-2000 is an extremely versatile groundwater-modeling code that has the capability to simulate transient groundwater flow in three dimensions subject to common boundary conditions used to represent hydrologic features such as streams, springs, drains, lakes, and evapotranspiration by phreatophytes.

Discussions of the numerical technique used in this study, the finite-difference method, and MODFLOW-2000 can be found in McDonald and Harbaugh (1988), Anderson and Woessner (1992), Harbaugh and others (2000), and Hill and others (2000).

Discretization

As mentioned above, numerical modeling requires that the model domain be divided into discrete regions or cells. For the model described in this report, the upper Klamath Basin was divided into cells with a lateral dimension of 2,500 ft by 2,500 ft aligned in a grid consisting of 285 east-west trending rows and 210 north-south trending columns (fig. 4). In the vertical dimension the model consists of three layers of varying thicknesses ranging from about 5 ft to 3,600 ft depending on topography and proximity to the edge of the model (fig. 5). The model layers are defined to correspond to hydrogeologic units where possible. The rectangular model grid comprises 179,550 cells of which 100,070 are active. Active cells are those for which groundwater flow is calculated and inactive cells are those that are outside of the modeled area.

Hydrologic characteristics were defined for each cell in the model using the Layer-Property Flow (LPF) package of MODFLOW. Layers were formulated as confined (meaning transmissivity remains constant) to help linearize the model and improve numerical stability.

Transient models require that time also be discretized into specific increments. MODFLOW requires two types of increments be defined, stress periods and time steps. A stress period is an interval over which specified boundary fluxes, such as recharge, and stresses (such as pumping) remain constant. Stress periods are subdivided into time steps. This further subdivision enables the model user to evaluate the timing of the hydrologic response to changes in stresses and also improves numerical stability. The upper Klamath Basin groundwater model has been set up to simulate quarterly stress periods from 1970 through 2004 water years. (Water years begin on October 1 and end on September 30. For example the 1970 water year starts October 1, 1969 and runs through September 30, 1970.) Each quarterly period is divided into five time steps.

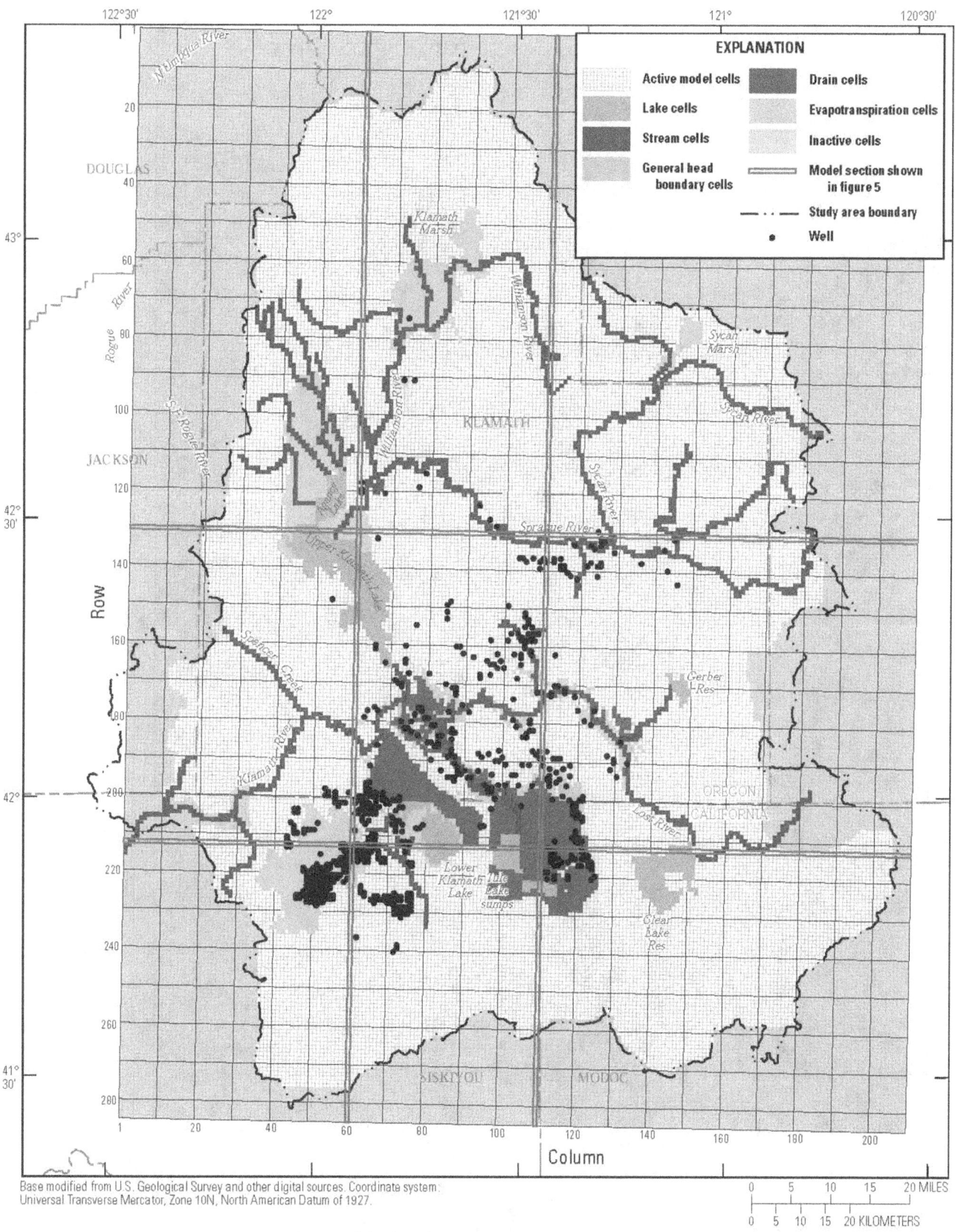

Figure 4. Upper Klamath Basin regional groundwater flow model grid and boundary conditions. Highlighted rows and columns correspond to sections shown in figure 5.

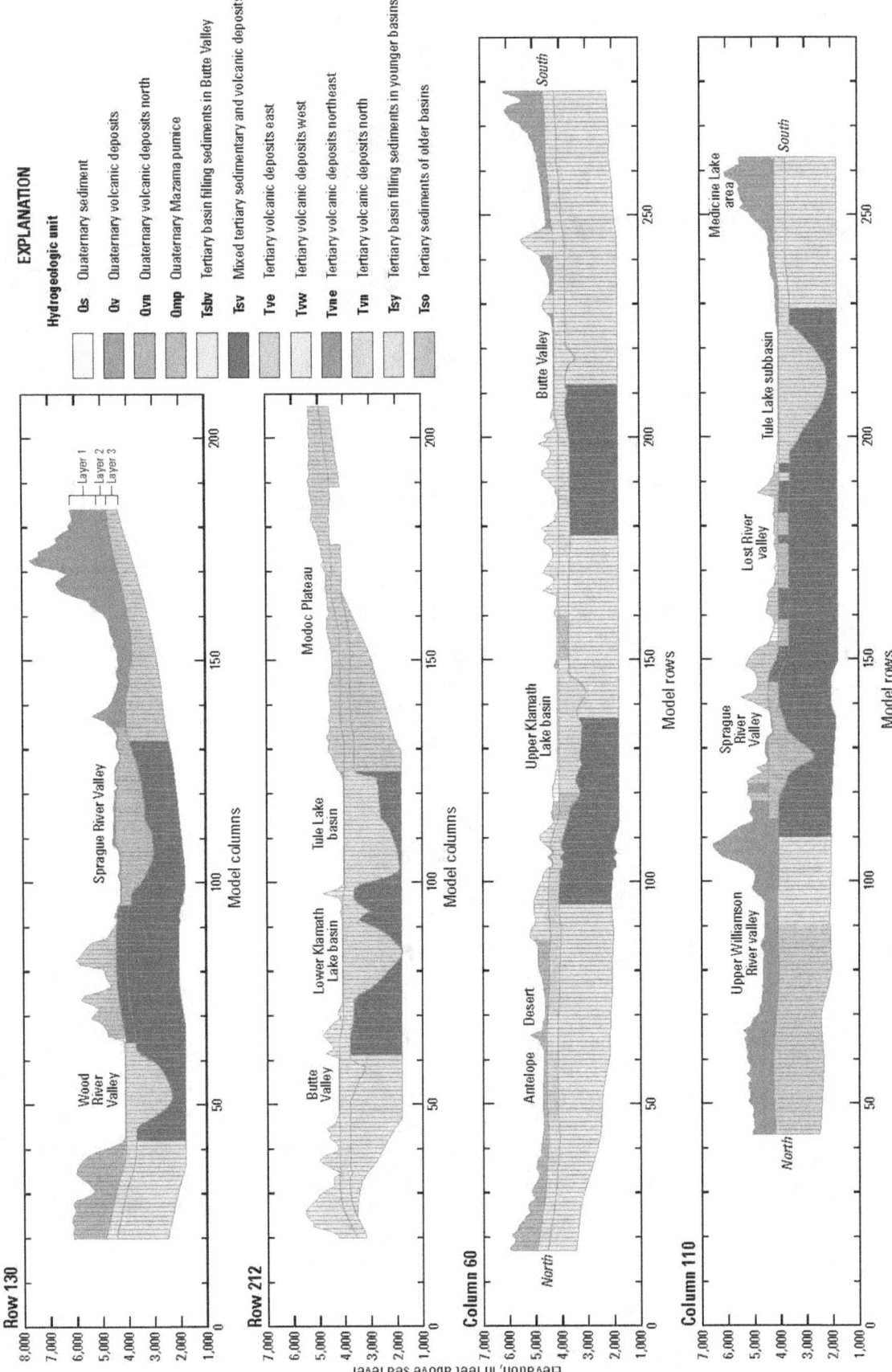

Figure 5. Model sections by hydrologeologic unit along selected rows and columns. Row and column locations are shown on figure 4.

Parameterization of Subsurface Hydraulic Characteristics

Simulating three-dimensional (3D) groundwater flow requires that the hydraulic properties of subsurface materials be represented in three dimensions. The hydraulic properties needed by the model are the hydraulic conductivity and the specific storage.

The distribution of hydraulic properties in the upper Klamath Basin groundwater model is based on geologic mapping, stratigraphic information from water wells, and geophysical data. Sufficient information does not exist to accurately represent the full complexity of the spatial distribution of hydraulic properties. Therefore, the spatial distribution of hydraulic properties is simplified by defining zones in the model that generally follow hydrogeologic units, and in which hydraulic properties are considered uniform. The hydraulic conductivity and specific storage values assigned to each zone are initially based on independent information such as aquifer tests, specific-capacity tests, or literature values appropriate to the dominant rock type, and then refined during model calibration.

The zonation of hydraulic properties used for the model is shown for each model layer in figure 6. The zones are primarily based on the dominant rock type in each region. Boundaries are based on the mapped geology and stratigraphic information from wells, as well as hydrologic information such as major changes in head gradients. The age of rock is also factored into the zonation. There is a general progression of the age of volcanic rocks at the surface across the upper Klamath Basin in which rocks generally increase in age from west to east. Quaternary volcanism is prominent along the western margin of the study area in the Cascade Range while the oldest (late Miocene) volcanic rocks most commonly are exposed in the northeast part of the basin (Sherrod and Pickthorn, 1992).

Zonation was based on generalized geology described in Gannett and others (2007) but refined by more detailed 1:24,000-scale geologic mapping done by the Oregon Department of Geology and Mineral Industries and compiled by Jenks (2007).

Quaternary Sediment (Qs)—This zone comprises the young surficial deposits in the uppermost parts of large sedimentary basins as well as alluvial deposits in stream valleys. These materials are usually shown as Quaternary sediment or alluvium on geologic maps. This zone occurs only in the uppermost layer of the model (layer 1). Layer 1 generally is about 45 ft thick in this zone.

Quaternary Volcanic Deposits (Qv)—This zone encompasses areas mapped as Quaternary volcanic deposits in the Cascade Range south of the Klamath River as well as volcanic deposits associated with Medicine Lake Volcano at the southern end of the model. The materials primarily are basaltic and andesitic lava flows, but there are vent deposits and pyroclastic deposits as well. The thickness of these deposits is not well known and has been inferred from surface exposures and topography. It is limited to model layer 1. Qv ranges in thickness from 27 to 3,400 ft, being thickest in the Cascade Range.

Quaternary Volcanic Deposits North (Qvn)—This zone corresponds to Quaternary volcanic deposits in the Cascade Range north of the Klamath River including the Crater Lake area. It is limited to model layer 1, and the lithology and thickness are similar to that of Qv.

Quaternary Mazama Pumice (Qmp)—This zone consists of pumice and ash deposited during the climactic eruption of Mount Mazama, which formed Crater Lake. The unit is limited to model layer 1 in the northern part of the model just east of Crater Lake in the Klamath Marsh area. Qmp ranges from 35 to 295 ft thick.

Tertiary Basin Filling Sediments in Butte Valley (Tsbv)—This zone corresponds to basin filling deposits in the Butte Valley structural basin. It consists of unconsolidated sedimentary deposits with grain sizes ranging from silt to gravel, and thicknesses ranging from 400 to 1,224 ft. This zone occurs only in model layer 2.

Tertiary Basin Filling Sediments in Younger Basins (Tsy)—This zone corresponds to the basin-filling deposits in the relatively young Lower Klamath Lake and Tule Lake structural basins. These are generally fine-grained sediments occurring mostly in model layer 2 but also in a small area of layer 1. This unit may include small sections of interbedded lava as well. The thickness in model layer 2 ranges from 400 to 2,611 ft. Although Tsy spans multiple model layers, it constitutes a single model parameter.

Tertiary Sediments of Older Basins (Tso)—This unit corresponds to lacustrine and fluvial sediments that occur mostly in the Lost and Sprague River subbasins. These deposits crop out in uplands separating the basins and underlay the stream valleys, including a large part of the lower Sprague River subbasin. The unit consists largely of fine-grained sediment, but the unit also contains coarser alluvium and hydrovolcanic deposits. The unit ranges in thickness from 30 to 1,488 ft in model layer 1 and 400 to 2,344 ft in layer 2. Although Tso spans multiple model layers it constitutes a single model parameter.

Mixed Tertiary Sedimentary and Volcanic Deposits— (Tsv) This zone includes areas where mapped sedimentary and volcanic deposits are complexly interbedded at a scale too fine to discriminate from well data at the scale of the model grid. It also includes deep regions not penetrated by wells in which the lithology must be inferred from overlying deposits and general understanding of the regional geology. This unit ranges in thickness from 400 ft to 1,129 ft in layer 2 and up to 2,000 ft in model layer 3. Separate model parameters are defined for Tsv in model layers 2 and 3 (Tsv2 and Tsv3).

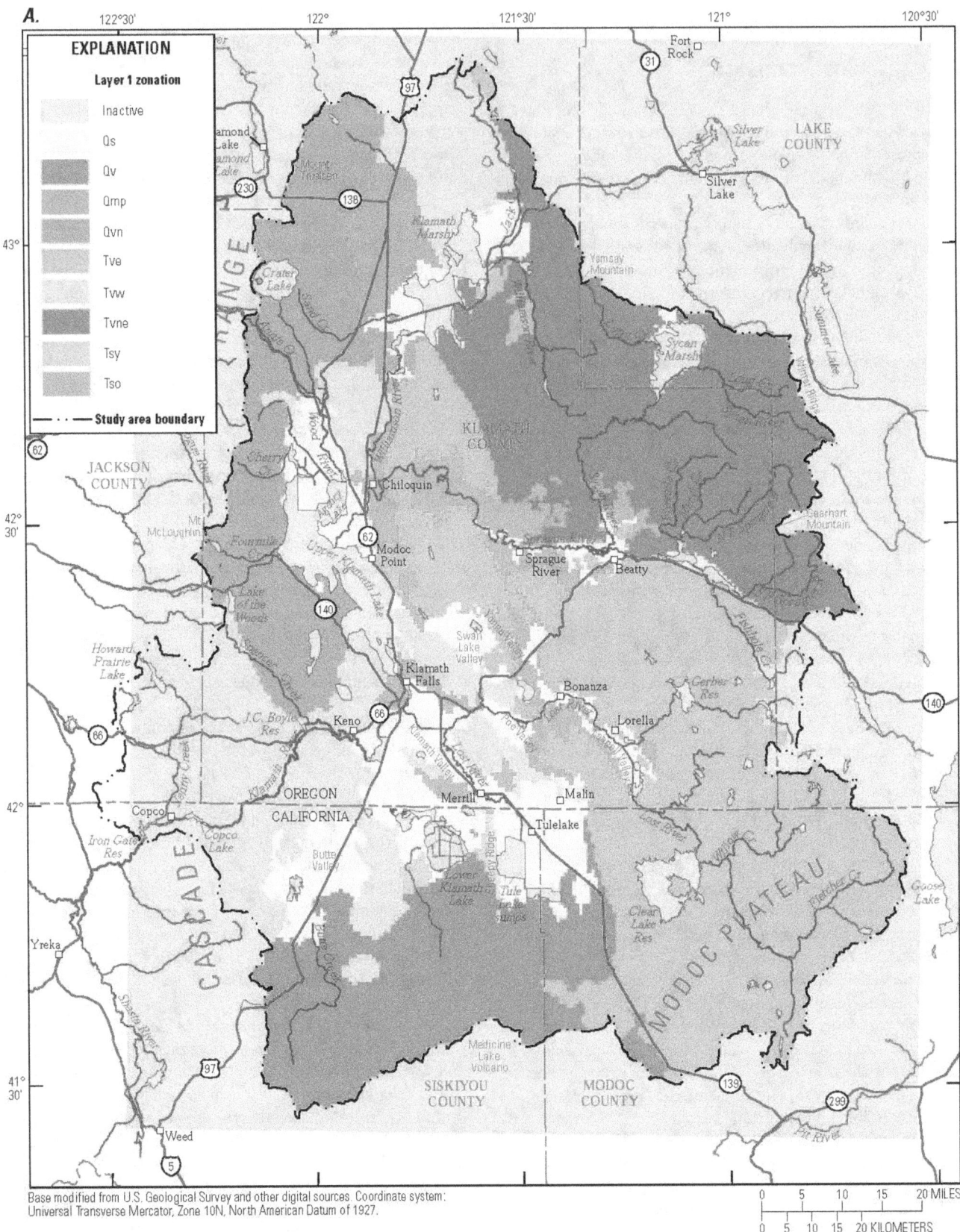

Figure 6.   Upper Klamath Basin, Oregon and California, regional groundwater flow model hydraulic-conductivity zonation. (*A*) Layer 1 zonation; (*B*) Layer 2 zonation; and (*C*) Layer 3 zonation. Hydrogeologic unit definitions: Qs, Quaternary sediment; Qv, Quaternary volcanic deposits; Qvn, Quaternary volcanic deposits north; Qmp, Quaternary Mazama pumice; Tvn, Tertiary volcanic deposits north; Tsbv, Tertiary sediments, Butte Valley; Tsy, Tertiary sediments younger basins; Tso, Tertiary sediments older basins; Tsv, Tertiary mixed sedimentary and volcanic deposits; Tvw, Tertiary volcanic rocks west; Tve, Tertiary volcanic rocks east; Tvne, Tertiary volcanic rocks northeast.

B.

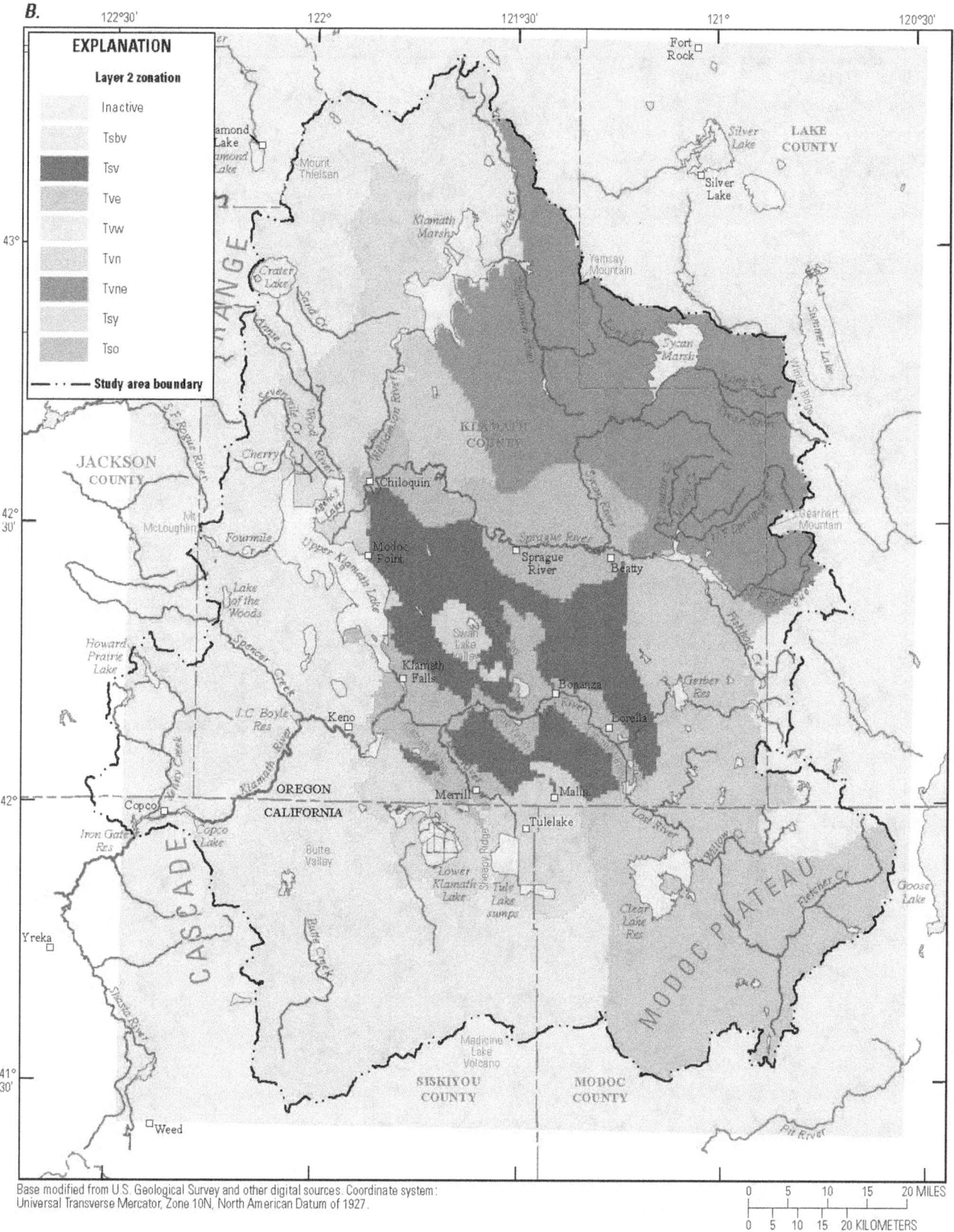

Figure 6.—Continued

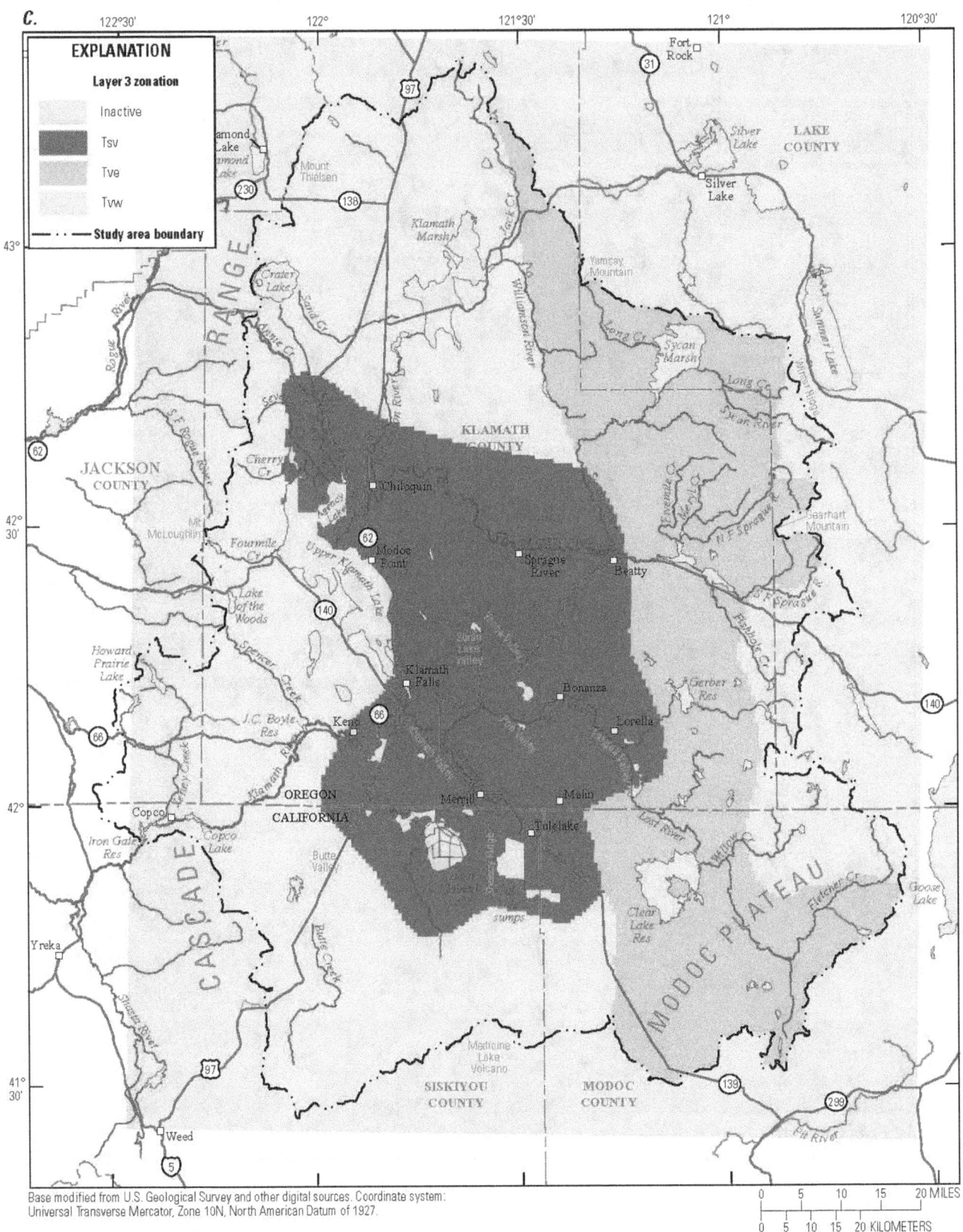

Figure 6.—Continued

Tertiary Volcanic Deposits North (Tvn)—This zone corresponds to the late Tertiary deposits from the Cascades east of Mount Mazama, generally underlying the pumice and ash deposits from the Crater Lake eruption. The exact lithology of materials in the zone is not well known but is assumed to be lava and volcaniclastic deposits. This zone occurs only in layer 2 and ranges from 400 to 654 ft thick.

Tertiary Volcanic Deposits West (Tvw)—This zone corresponds to volcanic deposits of mostly Pliocene age in the western part of the upper Klamath Basin. The lithology consists largely of basaltic lava flows and vent deposits, but is uncertain at depth due to the lack of well data. This zone includes parts of all three model layers and ranges in thickness from 25 to 3,200 ft. Separate model parameters are defined for Tvw in all three model layers (Tvw1, Tvw2, and Tvw3).

Tertiary Volcanic Deposits East (Tve)—This zone corresponds to volcanic deposits of Pliocene and Miocene age in the central and eastern parts of the upper Klamath Basin. The lithology consists largely of basaltic lava flows and vent deposits, but, as with other deposits in the area, is uncertain at depth due to the lack of well data. The zone occurs in all three model layers and ranges in thickness from 25 to 2,000 ft. Separate model parameters are defined for Tve in all three model layers (Tve1, Tve2, and Tve3).

Tertiary Volcanic Deposits Northeast (Tvne)—This zone corresponds to volcanic deposits of Pliocene and Miocene age in the northeastern part of the upper Klamath Basin. It consists largely of basaltic lava and vent deposits, but as with Tvw and Tve the lithology is uncertain at depth. This unit occurs only in model layers 1 and 2 with separate parameters defined for each layer (Tvne1 and Tvne2). The unit ranges in thickness from 27 to 3,350 ft.

Boundary Conditions

Boundary conditions define the manner in which water moves to or from the groundwater system. For example, the movement of water from the groundwater system to streams is one type of boundary condition. Boundary conditions vary in time and space. The types of boundary conditions used in the model are described in the following paragraphs.

Specified Flux Boundaries

Specified flux boundaries are locations where there is a specified flow of groundwater to or from the model. In some circumstances the flow may be specified as zero. Geologic contacts and drainage divides are examples of boundaries where the flow is specified as zero (no-flow boundaries). Specified flux boundary conditions with non-zero rates include recharge from precipitation and irrigation, and pumping.

No-Flow Boundaries

No-flow boundaries generally correspond to contacts with low-permeability rock or with groundwater divides across which groundwater flow is assumed negligible. Specific examples of no-flow boundary conditions on the west include the Jenny Creek and Howard Prairie Lake Areas where the model boundary corresponds with the contact between permeable high Cascade Range volcanic rocks and low permeability western Cascade Range deposits. The western model boundary from Mount McLoughlin north past Crater Lake corresponds to a groundwater divide. The eastern boundary of the model from the upper Lost River subbasin north to Winter Rim corresponds closely to the drainage divide and the contact with low-permeability middle to late Miocene deposits. Most divides between the upper Klamath Basin and adjacent basins are formulated as no-flow boundaries with the exception of two areas to the north and south, discussed below, that are formulated as general head boundaries. The bottom of the model (the base of layer 3) is also formulated as a no-flow boundary because it corresponds to the contact between the regional flow system and the underlying rock with very low permeability.

Recharge

The average rate of recharge during each stress period is specified for each active cell in the uppermost model layer. Recharge was initially estimated using the USGS Precipitation-Runoff Modeling System (PRMS), a watershed model that simulates the hydrologic processes affecting the routing, storage, and fate of water that falls as precipitation. Major processes simulated by PRMS include plant canopy interception, accumulation and melting of snow, evaporation, sublimation, accumulation and storage of soil moisture, transpiration by plants, direct runoff, routing of water through subsurface reservoirs to streams, and groundwater recharge. Complete descriptions of PRMS can be found in Leavesley and others (1983) and Markstrom and others (2008).

Watershed models like PRMS simulate runoff using daily values of precipitation, air temperature, and solar radiation. The watershed is divided into geographic subregions called hydrologic response units (HRUs). Spatially varying watershed characteristics such as elevation, soils, vegetation, and average precipitation are defined for each HRU. The responses of individual HRUs to meteorological inputs are integrated to determine the overall basin response.

Watershed models are calibrated by adjusting various parameters that represent key controls on the watershed response, such as the characteristics of soil, vegetation, and shallow aquifers, in order to simulate observed runoff as closely as possible. Proper calibration of a watershed model requires daily streamflow measurements, usually from gaging stations. Calibration is difficult in watersheds where streams are highly regulated or diverted, or where there is considerable groundwater flow to or from adjacent basins.

To estimate groundwater recharge, a single watershed model was developed for the Klamath Basin above Iron Gate Dam. The subsurface flow (interflow) and groundwater flow terms from the PRMS model were summed to estimate recharge. No groundwater sink term was used in the watershed model to maintain the basinwide water balance. Because the Klamath River is highly regulated above Iron Gate Dam, streamflow data were not suitable for a refined watershed model calibration. To provide information on key basin characteristics, a series of watershed models were calibrated for unregulated or minimally regulated basins at scales ranging from tens or hundreds of square miles to a thousand square miles. Representative parameter values were then applied to the basinwide model to estimate groundwater recharge. The resulting distribution of groundwater recharge from precipitation, shown as an average annual value, is shown in figure 7.

The spatial and temporal distribution of recharge is determined largely by precipitation, temperature, and topography, all of which are measured. The absolute volume of recharge, however, depends on quantities that are less well quantified, such as evapotranspiration.

The average annual subsurface flow (interflow) and groundwater recharge terms from the watershed model totaled nearly 3 million acre-ft/yr (1970–2004). The subsurface flow (interflow) term, however, represents relatively shallow rapid flow directly to streams that moves at timescales more similar to runoff than groundwater. This shallow rapid subsurface flow cannot be realistically simulated in a regional-scale groundwater model. During calibration, therefore, the net recharge values from the precipitation runoff model were adjusted downward to more accurately represent groundwater at the scales of interest and to improve the model fit with measured heads and groundwater discharge estimates. Recharge was adjusted independently in three zones in the model (fig. 7) corresponding to the Cascade Range, central low-elevation areas, and northeastern areas. The final average-annual precipitation recharge value was about 2.6 million acre-ft/yr. This is in reasonable agreement with the estimated average annual recharge figure of 2 million acre-ft/yr made by Gannett and others (2007) based on measurements and estimates of groundwater discharge.

Additional recharge from deep percolation of irrigation water is specified in areas irrigated with surface water. Deep percolation occurs when water is applied at a rate that exceeds the soil storage capacity and evapotranspiration. When this occurs, water moves through the soil to the shallow groundwater system. In most of the irrigated areas in the upper Klamath Basin, such water moves in the shallow subsurface to adjacent agricultural drains or streams.

Recharge from deep percolation was specified in the area of the Klamath Reclamation Project. The rate of deep percolation was estimated using the water balance for the Klamath Project by Burt and Freeman (2003). An analysis of their water balance suggests deep percolation rates of several tenths of a foot per year, although there is large uncertainty. A value of 1 ft/yr was applied throughout the Klamath Reclamation Project to account for deep percolation as well as some additional recharge from transmission losses, which are generally unmeasured but known to occur. The annual volume was proportioned to the quarterly stress periods to match total irrigation diversions. The total estimated recharge from all sources for each quarter is shown in figure 8.

Pumping

Groundwater pumping is specified for each stress period for cells in which the wells are located. Gannett and others (2007) described the methods used to determine groundwater pumping and provide estimated pumping for water years 2000 through 2004. Irrigation pumping in Oregon was estimated from water rights records and satellite imagery from 2000. Irrigation pumping in California was estimated from the California Department of Water Resources (CDWR) land use survey of 2000. These rates were used to estimate pumping back to 1970 taking into consideration variations in demand and the timing of groundwater development. An index of irrigation demand was created based on the consumptive use of the Klamath Reclamation Project as determined from Reclamation's monthly diversion and return flow records. For wells in Oregon, pumping was assumed not to occur in years earlier than the priority date of the water right.

Additional pumping for Reclamation's groundwater acquisition program in 2001 and pilot water bank in 2002 through 2004 was determined from flow-meter readings and well locations provided by Reclamation. The distribution of pilot water bank pumping for 2003 and 2004 is shown in figure 20 of Gannett and others (2007).

Municipal pumping in Oregon was based on water-use reporting data for recent years from the State of Oregon and was estimated for earlier years based on population data. Municipal pumping in California was based on population data. The spatial distribution of municipal pumping was based on known well locations.

The spatial distribution of irrigation pumping was determined differently for each State. For Oregon, irrigation pumping was tied to individual water rights, the vast majority of which have surveyed well locations. Pumping depths were determined from well logs that were tied to most water rights. Where well logs were not found, pumping depth was estimated from neighboring irrigation wells. For California,

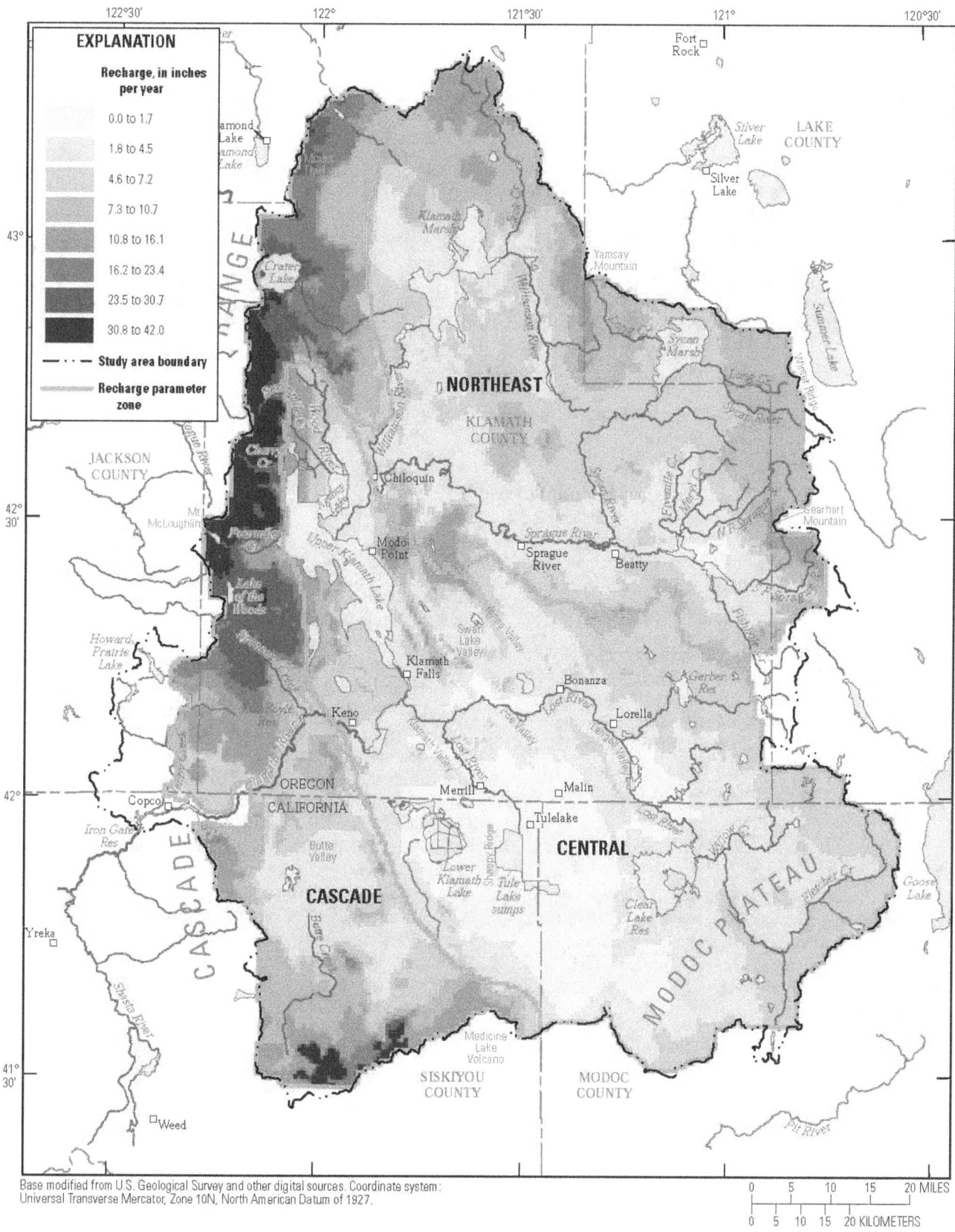

Figure 7. Estimated mean annual groundwater recharge from precipitation in the upper Klamath Basin, Oregon and California, 1970–2004, in inches, and recharge parameter zones.

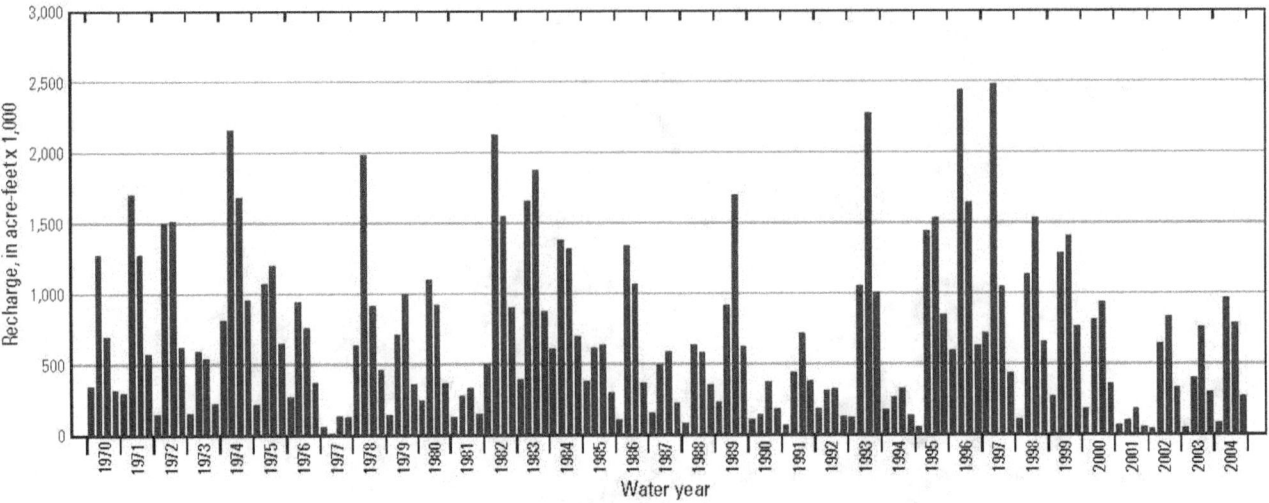

Figure 8. Estimated quarterly total groundwater recharge from all sources, upper Klamath Basin, Oregon and California, 1970 to 2004.

no information is available to correlate groundwater-irrigated fields to individual well locations or well logs. In this case, groundwater pumping was assumed to come from a well located at the center of the irrigated field. A pumping depth for each field was based on the depths of nearby irrigation wells determined from well logs. Generalizing the locations and depths of pumping in this manner is not considered problematic given the scale of the model. There are 906 wells in the model; 23 in model layer 1, 765 in layer 2, and 118 in layer 3. Total pumping in 2000 was about 160,000 acre-ft. Two percent of the pumping was from model layer 1, 81 percent from layer 2, and 17 percent from layer 3. The spatial distribution of pumped wells in 2000 is shown in figure 9, and quarterly pumping rates for 1970 to 2004 are shown in figure 10.

Head-Dependent Flux Boundaries

Head-dependent flux boundaries were used to simulate places or features where water moves to or from the groundwater system based on the hydraulic head in the aquifer. Head-dependent flux boundaries include streams, lakes, agricultural drains, some basin boundaries, and evapotranspiration.

Streams

The movement of groundwater to or from streams depends on the relation between the head in the aquifer (which can be thought of as the water-table elevation) and the stage of the stream. Where the head in the aquifer is higher than the stream stage, water will flow from the aquifer to the stream and the streams are said to be *gaining*. Such groundwater discharge usually occurs through springs or seepage through the streambed. Where the head in the aquifer is below the stream stage, water can leak from the stream to the aquifer, resulting in a *losing* stream. The rate of flow between the stream and the adjacent aquifer is proportional to the difference between the head in the aquifer and the stream stage, and the conductance of the streambed.

Streams were simulated using the MODFLOW stream package (STR6) (Prudic, 1989). All major streams and most large tributaries in the upper Klamath Basin are included in the model (fig. 4). Critical data requirements for this package are stream stage and streambed conductance. Stream stages were determined from a 10-meter digital elevation model and from 1:24,000 scale topographic maps. Stream stages were held constant during the simulations. Streambed conductance values were initially determined using streambed geometry estimated from 1:24,000 USGS quadrangle maps and streambed hydraulic conductivity set to match the surrounding bedrock. Streambed conductance was then adjusted during calibration.

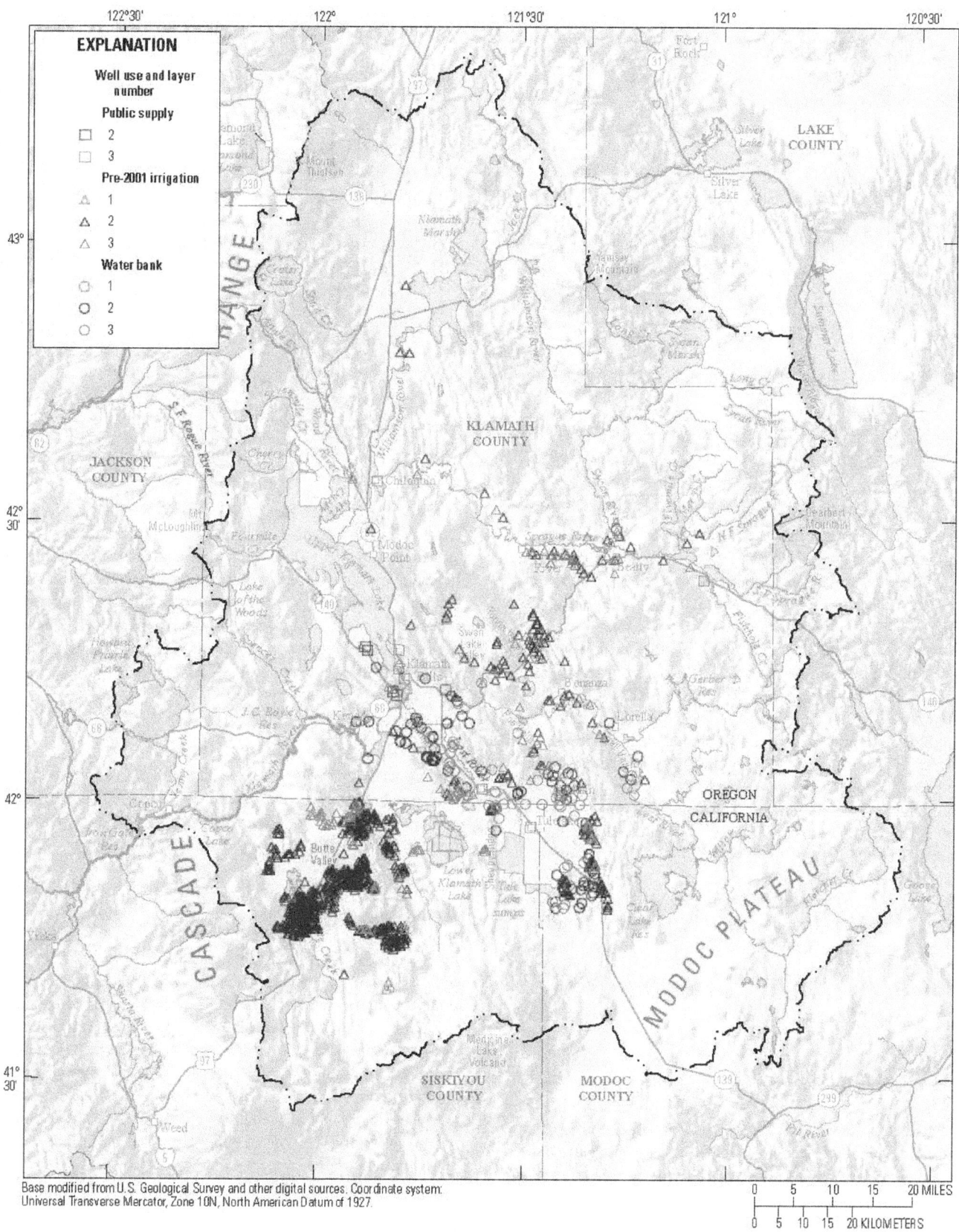

Figure 9. Distribution of pumping wells used in the upper Klamath Basin, Oregon and California, regional groundwater flow model, by model layer.

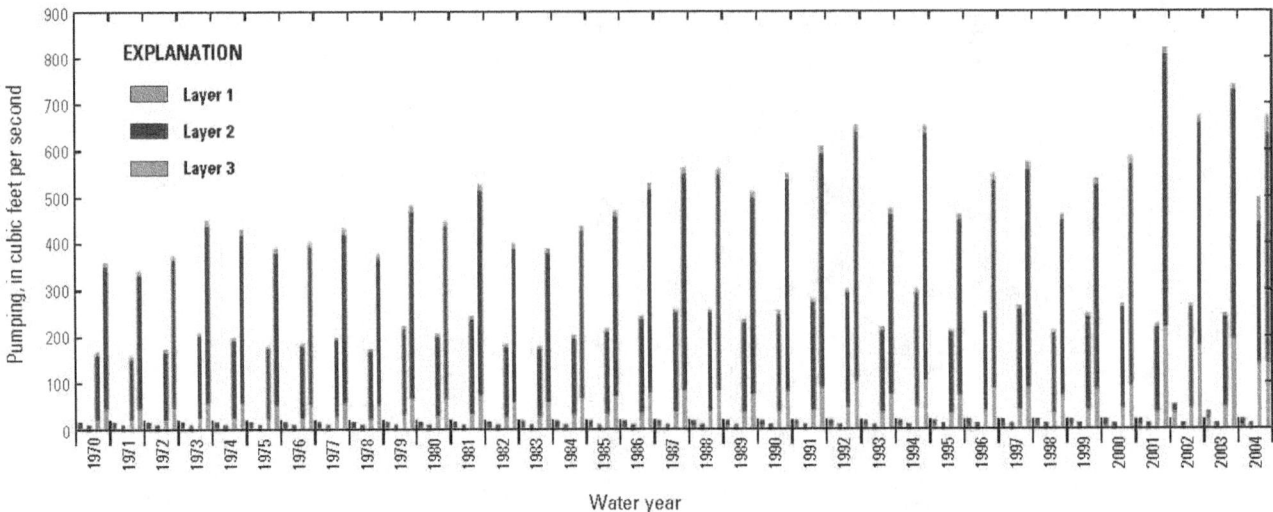

Figure 10.   Total estimated quarterly groundwater pumping in the upper Klamath Basin, Oregon and California, 1970–2004, by model layer.

The vast majority of streams in the upper Klamath Basin are either gaining or have very little net exchange with the groundwater system. The rates and distribution of groundwater discharge to streams in the basin are described in detail by Gannett and others (2007). Gaining streams are common because most major streams are in regional topographic lows that are areas of convergent groundwater flow. Losing streams are rare in the upper Klamath Basin, being restricted primarily to the pumice deposits in the upper Williamson Drainage immediately east of Crater Lake. Where geographic and hydrologic conditions are such that streams are above the water table and could potentially lose water to aquifers, the permeability of the streambed is commonly very low due to plugging by sediment from the stream.

To more accurately represent the actual conditions in the upper Klamath Basin and to greatly improve the numerical stability of the model, the streamflow routing package was set up to only allow water movement from the groundwater system to streams and not from streams to aquifers (effectively formulating streams as drains). This was accomplished by setting the streambed bottom elevation parameter (SBOT) to the stream stage. When head in the aquifer is above the stream stage (a gaining condition), the groundwater discharge to the stream is calculated as the product of the streambed conductance multiplied by the difference between the stream stage and the head in the aquifer. Stream bottom elevation is not involved in the calculation. When the head in the aquifer is below the stream stage (a losing condition), the stream leakage is calculated as the product of the streambed conductance multiplied by the difference between the stream

stage and the stream bottom elevation. When the stage and bottom elevation are the same, their difference becomes zero and calculated stream losses also become zero. The streams in the basin known to lose water to the groundwater system are generally small and do not represent a significant source of recharge. Modeling the streams in the manner described above more accurately represents actual conditions in the upper Klamath Basin.

Lakes

Lakes in the model were also simulated using head-dependent flux boundaries. The rate of groundwater discharge to lakes, or leakage from lakes to the groundwater system is proportional to the difference between the head in the aquifer and the stage of the lake, and a lakebed conductance term.

Because the lakes included in the model area are all artificially controlled, they were simulated using the MODFLOW reservoir package (RES1). Lakes simulated include Upper and Lower Klamath Lakes, the Tule Lake sumps, Gerber Reservoir, and Clear Lake. Principal data requirements for this package are the lake stage and lakebed conductance values. Lake stages for Upper Klamath Lake were based on historic measurements and were varied each stress period, ranging from 4,135.1 to 4,141.5 ft during the simulation period. Stage measurements for Clear Lake and Gerber Reservoir were available for only part of the simulation period, so stages were varied quarterly based on averages for the available periods of record. Stages for Clear Lake varied between 4,529.4 and 4,532.8 ft during the simulation period.

Stages for Gerber Reservoir varied between 4,814.4 and 4,830.2 ft. Because stage measurements were not available for the Tule Lake sumps or Lower Klamath Lake during the simulation period, their stages were fixed at long-term average values of 4,033 and 4,078 ft, respectively. Initial lakebed conductance values were set to reflect the permeability of the surrounding bedrock and geometry of the lakebed in the cell, and the values were adjusted during model calibration.

Drains

Agricultural drains in the model also were simulated as head-dependent flux boundaries. Groundwater discharges to drains whenever the hydraulic head in the aquifer (the water-table elevation) rises above the bottom of the drains. The rate of discharge is proportional to the difference between the water table and drain-bottom elevations, and a drain conductance term. Drains were simulated using the MODFLOW drain (DRN) package. The drain package differs from the stream package in that drains can only allow groundwater discharge and water cannot infiltrate to the groundwater system through drains. Principal input parameters for the drain package are the elevation of the drain bottoms and a drain conductance term. Drain bottoms were set at 10 ft below ground level, and initial drain conductance values (which represent the hydraulic conductivity of the soils around the drain) were based on hydraulic conductivity estimates of bedrock in the area and then adjusted during calibration. The distribution of drains in the model is based on the drains mapped on 1:24,000 scale topographic maps.

Evapotranspiration

Evapotranspiration from groundwater was also simulated as a head-dependent flux boundary. Most evapotranspiration in the basin involves water from the soil zone and not from the groundwater system. Water lost in this manner is returned to the atmosphere before it has a chance to become groundwater recharge. Water lost to evapotranspiration from the soil zone is calculated by the watershed model and is not available for recharge or runoff. Though most water lost through evapotranspiration in the basin comes from soil moisture, a small amount comes directly from groundwater. Evapotranspiration directly from groundwater occurs only in areas where the water table is close to land surface (within 10 ft or so) and where there are plants with roots that extend at least to the capillary fringe above the water table. Areas that meet these criteria in the upper Klamath Basin include the extensive wetlands in the areas of Sycan Marsh, Klamath Marsh, around Upper Klamath Lake, Lower Klamath Lake, parts of the Tule Lake subbasin, as well as agricultural lands in low-elevation areas throughout the basin (fig. 4).

Evapotranspiration directly from groundwater was simulated using the MODFLOW EVT package. With the EVT package, the rate of evapotranspiration by plants is inversely proportional to the depth of groundwater below land surface. When the water table is at the land surface, evapotranspiration occurs at a prescribed maximum rate. As the water table drops, the evapotranspiration is reduced linearly in response, becoming zero when the water table reaches the *extinction depth* at which plants can no longer extract water from the saturated zone. The extinction depth is a function of (a) the maximum rooting depth of plants and (b) soil properties.

Principal parameters for the EVT package are the maximum evapotranspiration rate and the extinction depth. The maximum evapotranspiration rate was based on the watershed model used to estimate groundwater recharge. The watershed model calculated a potential evapotranspiration rate (PET) based on meteorological factors such as solar radiation and temperature, as well as an actual evapotranspiration rate (AET) based on available moisture from precipitation. The difference between PET and AET represents the amount of potential demand not supplied by precipitation that could be provided by groundwater. This difference is used as the maximum evapotranspiration rate for the MODFLOW EVT package. This term varies seasonally and from year to year depending on meteorological conditions. The extinction depth was set to 10 ft. This value resulted in reasonable simulated water-table elevations and total evapotranspiration rates and is consistent with literature values (Canadell and others, 1996; Shah and others, 2007).

Interbasin Groundwater Flow

The final type of head-dependent flux boundary used in the model is the general head boundary. General head boundaries, simulated using the MODFLOW GHB package, allow movement of groundwater into or out of model cells based on the difference between the head in the cell and the head in an external source or sink (the boundary head). The rate of flow is proportional to the head difference between the cell and the source or sink. The proportionality is determined by a conductance term that incorporates hydraulic conductivity, cell geometry, and distance. General head boundaries were used to simulate interbasin groundwater flow between the upper Klamath Basin and the Deschutes Basin to the north and the Pit River Basin to the south (fig. 4). The external heads were set based on head measurements inside and outside of the model domain with the goal of representing the actual head gradient (to the extent known). The initial conductance values were set based on the hydraulic conductivity of model cells in the area and then adjusted during calibration.

Model Calibration

Model calibration is the process in which the model structure and model-parameter values are refined or adjusted within reasonable limits so that simulated conditions (heads and flows) match observed conditions as closely as possible. The terms *observed conditions* and *observations* as used herein refer to measured or estimated values of heads or flows derived independently of the model. The parameters that are adjusted during the calibration process include the hydraulic conductivity and specific storage values of the hydrogeologic units (as defined by the zones previously described), conductance terms for head-dependent flux boundaries, maximum ET rates, and recharge rates. Table 1 is a list of the calibration parameters for the upper Klamath Basin groundwater model and their final calibrated values. During calibration, the parameter values were adjusted within acceptable ranges to provide the best fit between observed hydraulic heads and fluxes and their simulated equivalents. Model calibration is a challenge because there is interaction between parameters such that the optimal value of one parameter is dependent on the values of other parameters. This section describes the overall calibration strategy, calibration data, specific approaches used, and the model fit.

For the transient calibration, boundary conditions such as recharge, groundwater pumping, maximum ET rates, and lake stage (in certain lakes) were varied by quarterly stress periods. Input datasets were developed for the period from 1970 through 2004. The model was calibrated for the period from 1989 through 2004. Beginning the simulation 19 years prior to the calibration period greatly reduces the influence of initial conditions on the calibration. A steady-state version of the model was used early in the calibration process to help develop the strategy for parameterizing aquifer properties. No final steady-state model calibration was developed, however, because of uncertainty regarding appropriate steady-state conditions.

Calibration Data

The model was calibrated using hydraulic-head measurements from water wells and estimates of groundwater discharge to streams derived from stream gage data and seepage runs. Time-series head data used for the calibration included 5,636 individual head observations from 663 wells. Of these, 444 wells had time series ranging from 2 to 64 observations. Head observations are assigned to a particular layer and X-Y location in the model grid and a time during the calibration period. During model calibration, observed heads are compared with simulated heads at the same X-Y location, model layer, and time. With very few exceptions, head measurements were made by the USGS, OWRD, or CDWR.

About 20 single measurements made by drillers or other third parties were used in the calibration dataset where no other data were available in a particular area.

Data are concentrated in populated parts of the basin and sparse in forested upland areas. In the vertical dimension, well depths are concentrated closer to land surface. Of the wells used for calibration, more than half are less than 300 ft deep, 80 percent are less than 500 ft deep, and 95 percent are less than 1,000 ft deep. The dataset includes only three wells with depths greater than 2,000 ft.

Time-series observations of groundwater discharge to streams or major springs (herein termed stream-flux observations) were available for 10 locations. These observations were derived from stream gage data and repeated discharge measurements of certain spring-fed streams (table 2). During calibration, stream-flux observations are compared to the summed groundwater fluxes discharging to groups of stream cells that best represent the stream network contributing to the field measurement. Gannett and others (2007) estimated long-term average groundwater discharge to 52 stream reaches or spring complexes (table 3). These data were not used for transient model calibration, but were used for preliminary steady-state model calibration and for evaluating the spatial distribution of groundwater discharge simulated by the transient model.

Calibration Methods

The model was calibrated using parameter estimation, a technique that uses computational methods to determine the set of parameter values that provides the best fit between observed and simulated dependent (system) variables, which in the case of this model are heads and stream fluxes. Parameter estimation requires some mathematical measure of the goodness of model fit, referred to as an objective function. For this model, a weighted sum-of-squares objective function, defined as $S(\underline{b})$, was used (from Hill and Tiedeman, 2007, p. 27):

$$S(\underline{b}) = \sum_{i=1}^{ND} w_i \left[y_i - y'_i(\underline{b}) \right]^2, \tag{2}$$

where

$\underline{b}$ is a vector containing values of each of the parameters being estimated,

ND is the number of observations,

y_i is the ith observation being matched by the regression,

$y'_i(\underline{b})$ is the simulated value corresponding to the ith observation, which is a function of $\underline{b}$, and

w_i is the weight for ith observation.

Table 1. Calibration parameters and their final values for the upper Klamath Basin, Oregon and California, regional groundwater flow model.

[**Region or process:** Layer refers to model layer. **Abbreviations:** ft^{-1}, 1/foot, ft/s, foot per second]

Parameter type	Region or process	Label	Value	Unit
Hydraulic conductivity	Quaternary sediment	HK_QS	5.80E–3	ft/s
Vertical anisotropy	Quaternary sediment	VANI_QS	1.76E+1	dimensionless
Specific storage	Quaternary sediment	SS_QS	1.00E–3	ft^{-1}
Hydraulic conductivity	Quaternary volcanic deposits	HK_QV	5.90E–6	ft/s
Vertical anisotropy	Quaternary volcanic deposits	VANI_QV	1.00E+2	dimensionless
Specific storage	Quaternary volcanic deposits	SS_QV	5.28E–4	ft^{-1}
Hydraulic conductivity	Quaternary volcanic deposits north	HK_QVN	3.99E–5	ft/s
Vertical anisotropy	Quaternary volcanic deposits north	VANI_QVN	1.00E+3	dimensionless
Specific storage	Quaternary volcanic deposits north	SS_QVN	2.00E–6	ft^{-1}
Hydraulic conductivity	Quaternary Mazama pumice	HK_QMP	1.15E–2	ft/s
Vertical anisotropy	Quaternary Mazama pumice	VANI_QMP	1.00E+3	dimensionless
Specific storage	Quaternary Mazama pumice	SS_QMP	5.00E–4	ft^{-1}
Hydraulic conductivity	Tertiary volcanic deposits north layer 2	HK_TVN2	9.31E–4	ft/s
Vertical anisotropy	Tertiary volcanic deposits north layer 2	VANI_TVN2	1.00E+1	dimensionless
Specific storage	Tertiary volcanic deposits north layer 2	SS_TVN2	7.29E–7	ft^{-1}
Hydraulic conductivity	Tertiary sediments—Butte Valley	HK_TSBV	3.50E–3	ft/s
Vertical anisotropy	Tertiary sediments—Butte Valley	VANI_TSBV	8.17E+1	dimensionless
Specific storage	Tertiary sediments—Butte Valley	SS_TSBV	2.57E–4	ft^{-1}
Hydraulic conductivity	Tertiary sediments—Younger basins	HK_TSY	2.90E–4	ft/s
Vertical anisotropy	Tertiary sediments—Younger basins	VANI_TSY	2.50E+2	dimensionless
Specific storage	Tertiary sediments—Younger basins	SS_TSY	7.00E–5	ft^{-1}
Hydraulic conductivity	Tertiary sediments—Older basins	HK_TSO	4.13E–4	ft/s
Vertical anisotropy	Tertiary sediments—Older basins	VANI_TSO	1.00E+1	dimensionless
Specific storage	Tertiary sediments—Older basins	SS_TSO	9.24E–5	ft^{-1}
Hydraulic conductivity	Tertiary mixed sedimentary and volcanic deposits layer 2.	HK_TSV2	1.06E–5	ft/s
Vertical anisotropy	Tertiary mixed sedimentary and volcanic deposits layer 2.	VANI_TSV2	1.66E+2	dimensionless
Specific storage	Tertiary mixed sedimentary and volcanic deposits layer 2.	SS_TSV2	1.00E–5	ft^{-1}
Hydraulic conductivity	Tertiary mixed sedimentary and volcanic deposits layer 3.	HK_TSV3	1.35E–5	ft/s
Vertical anisotropy	Tertiary mixed sedimentary and volcanic deposits layer 3.	VANI_TSV3	1.00E+1	dimensionless
Specific storage	Tertiary mixed sedimentary and volcanic deposits layer 3.	SS_TSV3	1.60E–5	ft^{-1}
Hydraulic conductivity	Tertiary volcanic rocks west layer 1	HK_TVW1	3.12E–4	ft/s

Table 1. Calibration parameters and their final values for the upper Klamath Basin, Oregon and California, regional groundwater flow model.—Continued

[**Region or process:** Layer refers to model layer. **Abbreviations:** ft^{-1}, 1/foot, ft/s, foot per second]

Parameter type	Region or process	Label	Value	Unit
Vertical anisotropy	Tertiary volcanic rocks west layer 1	VANI_TVW1	1.00E+3	dimensionless
Specific storage	Tertiary volcanic rocks west layer 1	SS_TVW1	1.22E–6	ft^{-1}
Hydraulic conductivity	Tertiary volcanic rocks west layer 2	HK_TVW2	1.16E–4	ft/s
Vertical anisotropy	Tertiary volcanic rocks west layer 2	VANI_TVW2	1.00E+3	dimensionless
Specific storage	Tertiary volcanic rocks west layer 2	SS_TVW2	2.20E–7	ft^{-1}
Hydraulic conductivity	Tertiary volcanic rocks west layer 3	HK_TVW3	5.79E–4	ft/s
Vertical anisotropy	Tertiary volcanic rocks west layer 3	VANI_TVW3	2.18E+1	dimensionless
Specific storage	Tertiary volcanic rocks west layer 3	SS_TVW3	1.00E–7	ft^{-1}
Hydraulic conductivity	Tertiary volcanic rocks east layer 1	HK_TVE1	1.00E–5	ft/s
Vertical anisotropy	Tertiary volcanic rocks east layer 1	VANI_TVE1	1.00E+3	dimensionless
Specific storage	Tertiary volcanic rocks east layer 1	SS_TVE1	1.81E–4	ft^{-1}
Hydraulic conductivity	Tertiary volcanic rocks east layer 2	HK_TVE2	2.94E–5	ft/s
Vertical anisotropy	Tertiary volcanic rocks east layer 2	VANI_TVE2	1.00E+1	dimensionless
Specific storage	Tertiary volcanic rocks east layer 2	SS_TVE2	1.91E–6	ft^{-1}
Hydraulic conductivity	Tertiary volcanic rocks east layer 3	HK_TVE3	3.56E–5	ft/s
Vertical anisotropy	Tertiary volcanic rocks east layer 3	VANI_TVE3	1.00E+1	dimensionless
Specific storage	Tertiary volcanic rocks east layer 3	SS_TVE3	1.00E–7	ft^{-1}
Hydraulic conductivity	Tertiary volcanic rocks northeast layer 1	HK_TVNE1	5.53E–5	ft/s
Vertical anisotropy	Tertiary volcanic rocks northeast layer 1	VANI_TVNE1	1.00E+3	dimensionless
Specific storage	Tertiary volcanic rocks northeast layer 1	SS_TVNE1	2.42E–5	ft^{-1}
Hydraulic conductivity	Tertiary volcanic rocks northeast layer 2	HK_TVNE2	3.02E–5	ft/s
Vertical anisotropy	Tertiary volcanic rocks northeast layer 2	VANI_TVNE2	1.00E+1	dimensionless
Specific storage	Tertiary volcanic rocks northeast layer 2	SS_TVNE2	7.49E–7	ft^{-1}
Multiplier	General head boundary conductance—Southern boundary.	SOUTHBOUND	1.10E+2	dimensionless
Multiplier	General head boundary conductance—Northern boundary.	NORTHBOUND	7.96E+1	dimensionless
Multiplier	Maximum evapotranspiration rate—Non-irrigated lowlands.	ET_ACTV	1.00E+0	dimensionless
Multiplier	Maximum evapotranspiration rate—Irrigated areas.	ET_IRR	1.00E+0	dimensionless
Multiplier	Drain conductance	DRAINCOND	5.00E+0	dimensionless
Multiplier	Recharge—Cascade	RCH_CSCADE	7.00E–1	dimensionless
Multiplier	Recharge—Central	RCH_CNTRL	5.00E–1	dimensionless
Multiplier	Recharge—Northeast	RCH_NE	1.50E+0	dimensionless
Multiplier	Streambed conductance	STREAMCOND	5.00E+0	dimensionless
Multiplier	Resevoir bed conductance	RESCOND	1.00E+1	dimensionless

Table 2.   Transient stream-flux observation locations used in calibration of the upper Klamath Basin, Oregon and California, regional groundwater flow model.

[**Abbreviations:** USFS, U.S. Forest Service; USGS, U.S. Geological Survey; mi, mile; RM, river mile]

Observation name	Stream(s)	Source of groundwater discharge (baseflow) estimate
BON	Lost River near Bonanza Spring.	Six spring discharge estimates made from synoptic streamflow measurements.
SPR	Upper Sprague River—Headwaters to the gage near Beatty.	September mean flows at the USGS gage near Beatty, 1969–2004.
UWIL	Upper Williamson River—Entire drainage above the gage near Sheep Creek (RM 67.7).	September mean flows at the Oregon Water Resources Department gage below Sheep Creek, 1974–2003.
LWIL	Lower Williamson River—Williamson River between the gage near Klamath Agency (the Kirk gage) and the gage below the Sprague River; the Sprague River below the gage near Chiloquin, and Spring Creek.	Monthly estimates of groundwater inflow based on differences in flow at stream gages on the Williamson near the Klamath Agency and below the Sprague River, and on the Sprague River near Chiloquin, 1969–2004.
SAND	Sand Creek—Entire drainage above RM 5.8.	September mean flows at RM 5.8 (data from USFS gage about 4.5 mi west of U.S. Highway 97), 1993–2002.
WOOD	Wood River headwaters springs.	122 instantaneous measurements made at Dixon Road.
SEV	Sevenmile Creek—Entire drainage above RM 17.	September mean flows at RM 17 (data from USFS gage about 1 mi north of USFS Road 3300), 1993–2002.
CHER	Cherry Creek—Entire drainage above Westside Road.	September mean flows (data from USFS gage about 0.5 mi west of Westside Road) 1993–2002.
KLAM	Klamath River between Keno and the gage below J.C. Boyle Powerplant.	August accretions based on USGS gages at Keno and below J.C. Boyle Powerplant, and reservoir content data from PacifiCorp.
SPEN	Spencer Creek—Entire drainage.	September mean flows from USFS gage about 1 mi below Buck Lake, 1993–98.

Table 3.   Long-term average stream-flux observations used in calibration of the upper Klamath Basin, Oregon and California, regional groundwater flow model and their simulated equivalents from the transient model.

[Discharge values are in cubic feet per second. **Abbreviations:** mi, mile; RM, river mile;, n/a, not applicable]

Observation name	Observation location	Estimated groundwater discharge	Estimated 2 sigma	Confidence intervals		Simulated groundwater discharge
				Lower	Upper	
WILLOWCLEAR	Willow Creek above Clear Lake	28	27.5	0.5	55.5	11.99
LR–CLR–MAL	Lost River, Clear Lake (RM 78) to Malone Dam (RM 64.5).	2	1.0	1.0	3.0	7.05
LR–MAL–GFT	Lost River, Malone Dam (RM 64.5) to Gift Road (RM 58).	11	5.4	5.6	16.4	18.00
LR–GFT–KEL	Lost River, Gift Rd (RM 58) to Keller Bridge (RM 50.3).	16	7.8	8.2	23.8	19.69
GERBERMILLER	Miller Creek below Gerber Reservoir	4	2.0	2.0	6.0	5.67
LR–KEL–BON	Lost River, Keller Bridge (RM 50.3) to Bonanza (RM 45.1).	9	2.6	6.4	11.6	15.46
LR–BON–HARB	Lost River, Bonanza (RM 45.1) to Harpold Bridge (RM 43.9).	61	18.0	43.0	79.0	8.03
LR–HARB–KIRS	Lost River, Harpold Bridge (RM 43.9) to Kirsh Bridge (RM 34.8).	13	6.4	6.6	19.4	3.61
LR–KIRS–STEV	Lost River, Kirsh Bridge (RM 34.8) to Stevenson Park (RM 30.5).	19.7	9.7	10.0	29.4	0.20
SFSR–HW–PIC	South Fork Sprague River above picnic area (RM 10.2).	19	5.6	13.4	24.6	85.18
NFSR–HW–M1.5	North Fork Sprague River above RM 1.5.	59	17.3	41.7	76.3	102.71
FIVE–HW–RM4	Fivemile Creek above RM 4, and Meryl Creek above RM 1.5.	33	9.7	23.3	42.7	49.48
SR–SFPIC–RR	Sprague River, South Fork picnic area (South Fork RM 10.2) to railroad bridge (RM 74.2).	86	25.3	60.7	111.3	111.19
SR–RR–WDOG	Sprague River, railroad bridge (RM 74.2) to Watchdog Butte (RM 49).	28	13.7	14.3	41.7	127.81
SYCAN–HW–MSH	Sycan River above Sycan Marsh.	8	1.6	6.4	9.6	89.79
LONG–UPPR	Long Creek above T31S, R12E, Sec 5	9	0.9	8.1	9.9	0.00
LONG–LOWER	Long Creek below T31S, R12E, Sec 5	7	3.4	3.6	10.4	29.44
SYCAN–MSH–TO	Sycan River, Marsh (RM 35) to Torrent Spring (RM 30).	0	n/a	n/a	n/a	6.51
TORRENT–SPR	Sycan River near Torrent Spring (RM 30 to RM 24).	12	3.5	8.5	15.5	9.71
SYCAN–TOR–BL	Sycan River, Torrent Spring (RM 24) to Blue Creek (RM 11).	0	n/a	n/a	n/a	1.39
SYCAN–BLU–GA	Sycan River, Blue Creek (RM 11) to Gage (RM 3).	9	2.6	6.4	11.6	76.48
WHISKEY	Whiskey Creek, headwaters to RM 1	13	6.4	6.6	19.4	20.87
SR–WDOG–TRT	Sprague River, Watchdog Butte (RM 49) to below Trout Creek (RM 37).	0	n/a	n/a	n/a	20.77
SR–TRT–LONE	Sprague River, below Trout Creek (RM 37) to Lone Pine (RM 33).	0	n/a	n/a	n/a	7.80
SR–LONE–BRAY	Sprague River, Lone Pine (RM 33) to Braymill (RM 10.1).	73	21.5	51.5	94.5	83.91

Table 3. Long-term average stream-flux observations used in calibration of the upper Klamath Basin, Oregon and California, regional groundwater flow model and their simulated equivalents from the transient model.—Continued

[Discharge values are in cubic feet per second. **Abbreviations:** mi, mile; RM, river mile;, n/a, not applicable]

Observation name	Observation location	Estimated groundwater discharge	Estimated 2 sigma	Confidence intervals		Simulated groundwater discharge
				Lower	Upper	
SR–BRAY–CHIL	Spague River, Braymill (RM 10.1) to Chiloquin (RM 5.2).	0	n/a	n/a	n/a	0.11
WR–HW–RM81	Williamson River, headwaters to RM 81.	16	3.1	12.9	19.1	3.38
WR–RM81–SHP	Williamson River, RM 81 to Sheep Creek (RM 67.7).	38	7.5	30.5	45.5	45.33
WR–SHP–CHO	Williamson River, Sheep Creek (RM 67.7) to Cholo Ditch (RM 58.3).	18	8.8	9.2	26.8	21.55
WR–CHO–KIRK	Williamson River, Cholo Ditch (RM 58.3) to Kirk (RM 27).	50	24.5	25.5	74.5	160.68
LENZ–BIGSPR	Big Springs Creek to Lenz Creek	34	16.6	17.4	50.6	87.04
SAND	Sand Creek above RM 5.8	44	12.9	31.1	56.9	0.02
WR–CAL–GAG	Williamson River RM 22.2 to RM 19.7.	22	6.5	15.5	28.5	53.04
WR–19.7–17.5	Williamson River RM 19.7 to RM 17.5.	26	13	13.3	38.7	10.84
SPRING	Spring Creek just above mouth	300	29	270.6	329.4	289.85
SUN	Sun Creek above RM 5.0	13	3	10.5	15.5	0.43
ANNIE	Annie Creek above RM 5.7	56	11	45.1	66.9	21.59
WOOD–HW–DIX	Wood River, headwaters to Dixon Road (RM 16.2).	250	25	225.5	274.5	273.67
WOOD(FORT)	Fort Creek, headwaters to RM 1.0	84	8	75.8	92.2	40.29
CROOKED	Crooked Creek, headwaters to RM 5.4	43	8	34.6	51.4	9.50
TECUMSEH	Tecumseh Springs	27	8	19.1	34.9	2.66
AGENCY	Agency Creek, headwaters to mouth	21	6	14.8	27.2	0.19
7MI–UPPR	Sevenmile Creek, headwaters to RM 17.	18	4	14.5	21.5	18.28
7MI–LOWR	Sevenmile Creek, RM 17 to Sevenmile Road.	60	12	48.3	71.7	47.42
CHERRY	Cherry Creek, headwaters to Westside Road.	12	2	9.6	14.4	7.07
KENO–BOYLE	Klamath River, Keno (RM 230.5) to below JC Boyle Dam (RM 218.1).	190	37	152.7	227.3	184.93
SPENCER	Spencer Creek, headwaters to mouth	27	8	19.1	34.9	0.00
BOYLE–IGD	Klamath River, below JC Boyle Dam (RM 218.1) to below Iron Gate Dam (RM 189.5).	92	27	65.0	119.0	228.17
FALL	Fall Creek, headwaters to mouth	45	9	36.2	53.8	86.92
WILLOW	Willow Creek, 3 mi below headwaters spring.	5.8	3	3.0	8.6	3.08
COTTONWOOD	Cottonwood Creek, headwaters springs.	10	5	5.1	14.9	18.05
SHEEPY	Sheepy Creek, about 1 mi below headwaters springs.	16	8	8.2	23.8	14.95

As model fit is improved, the differences between the observed and simulated values $(y_i - y'_i)$, referred to as the residuals, become smaller, resulting in a smaller value of $S(\underline{b})$. Therefore, a lower value of the objective function indicates a better model fit.

In parameter estimation, nonlinear regression is used to determine the set of parameter values that provides the lowest value of $S(\underline{b})$, and presumably the best possible fit for a given model. Discussions of applicable nonlinear regression techniques can be found in Hill (1992) and Hill and Tiedeman (2007).

It is important to differentiate between parameters and actual model inputs. Many parameters correspond directly to model input values. For example, the single hydraulic conductivity value for a particular hydrogeological zone can be defined as a model parameter. In other cases, such as with recharge rates and stream-conductance terms, the actual model input values vary from cell to cell, resulting in far too many different input values for each to be defined as a separate parameter. Where inputs for a particular boundary condition are spatially or temporally variable, they are often grouped together, and the initial values adjusted in unison by a single parameter that is usually formulated as a multiplier. For example, the streambed-conductance parameter is a single value by which the initial conductance values, estimated from stream geometry and hydraulic conductivity of surrounding materials, are multiplied.

There were 64 parameters used for model calibration (table 1). Of these, 54 correspond to hydraulic conductivity, specific storage, and vertical anisotropy terms for 18 hydrogeologic unit zones previously described. Five of the parameters are multipliers applied to conductance terms for drains, streams, reservoir bottoms, and north and south general head boundaries. The remaining five parameters are multipliers applied to specified fluxes including maximum ET rates in irrigated and non-irrigated areas, and recharge in three zones.

Observation Weighting

Observations are weighted to control their relative influence on the objective function. A principal reason for weighting observations is to account for differences in measurement error or other uncertainty between observations. Weights are calculated as the inverse of the variance. In this way, observations with large error or uncertainty will have less influence on the objective function than those with very low error. The weighted squared residuals in equation (2) also have the advantage of being dimensionless, making it possible to compare (and sum) observations of different types.

The weighting of head observations used for calibration of the upper Klamath Basin groundwater flow model was based initially on estimates of measurement error. The largest source of error for most head observations was considered to be the determination of well elevations from topographic maps. Weights for such observations were based on an assumed confidence interval of plus or minus one contour interval of the topographic map used. Elevations for approximately 260 wells, mostly on the very flat floors of interior subbasins, were determined using survey-grade differential GPS measurements with an estimated error in the centimeter range. Weights based on this small measurement error resulted in very large weights that dominated the objective function. In order to prevent these wells, which are geographically clustered, from having undue influence on the model calibration, weights were based on an assumed standard deviation of error of 2 ft.

Initial parameter-estimation runs using the weighting procedure described above indicated the sum of weighted squared residuals was dominated by the head observations, resulting in an inadequate fit to discharge observations. This results from the fact that the number of head observations is seventeen times the number of discharge observations, and the weights assigned to head observations are generally much larger than the weights assigned to discharge observations. Additionally, the weights described above account for measurement errors only and do not account for model errors. Model errors are those errors that could be eliminated or reduced by changes in the model (Hill and Tiedeman, 2007, p. 300), such as finer discretization and parameter zonation. To create a set of weights that represents both measurement and model error, the weights for head observations were reduced by adding 10 ft to the standard deviations of head-measurement errors. These adjustments improved overall model fit without substantially degrading the fit to groundwater-head observations.

Weights for transient stream-flux observations were based on the error estimates commonly associated with streamflow measurements or, in the case of gaging-station data, as indicated in the published streamflow records. Initial calibration runs indicated the need to decrease weights on discharge observations for Cherry Creek (10 measurements) and Spencer Creek (6 measurements). These observations dominated the residuals for discharge observations and contributed a substantial portion of the total weighted sum of squares. To represent the errors associated with these observations, their initial weights were decreased on average by a factor of five.

Sensitivities

Model sensitivity describes the relation between dependent variables and parameter values. In this application, sensitivities are calculated as the derivative of the simulated equivalent of an observation with respect to a particular parameter value:

$$\left(\frac{\partial y'_i}{\partial b_j} \right)\Bigg|_{\underline{b}} , \qquad (3)$$

where

y'_i is the simulated value corresponding to the ith observation, and

b_j is the jth parameter.

The $\underline{b}$ notation indicates that the sensitivity is specific to a particular set of parameter values. This is needed for nonlinear models (such as the model described here) in which sensitivities are dependent on specific parameter values.

Because observations and parameters can both have a variety of units, it can be difficult to make comparisons between different observations. For example, observations may be in feet of elevation (for heads) or cubic feet per second (for flow), and parameter values may be in feet per second, inverse feet, or dimensionless. In MODFLOW, sensitivities are multiplied by the parameter value and the square root of the observation weight to calculate a dimensionless scaled sensitivity (ss_{ij}):

$$ss_{ij} = \left(\frac{\partial y'_i}{\partial b_j} \right) b_j w_i^{1/2} . \qquad (4)$$

Dimensionless scaled sensitivities can be used to compare the relative importance of particular observations to particular parameters. A measure of the total information about a particular parameter provided by all of the observations is provided by the composite scaled sensitivity (css_j):

$$css_j = \left[\sum_{i=1}^{ND} \left(ss_{ij} \right)^2 \Bigg|_{\underline{b}} / (ND) \right]^{1/2} . \qquad (5)$$

Generally speaking, regression techniques have more difficulty estimating values for parameters with low composite scaled sensitivities, and the uncertainties associated with such parameters are large relative to more sensitive parameters. Avoiding insensitive parameters is often difficult, however, due to poor spatial distribution of data. In cases for which the regression process failed to estimate parameter values because of low sensitivity, parameter values were given fixed (nonchanging) values. These fixed values were chosen on the basis of independent estimates.

The parameter-estimation software used for model calibration, PEST (Doherty, 2010), lets the user specify an allowable range of parameter values. As the regression process changes a parameter value, it will stop at this limit. Final parameter values at the limits of the allowable range indicate that the regression process may have ultimately resulted in a final value outside the range, a situation that often results from low parameter sensitivity.

Final Parameter Values

The final parameter values are given in table 1 and shown graphically in figure 11. Of the 64 parameters, 50 were determined by parameter estimation and 14 were fixed. Of the 50 parameters estimated, final values for 23 of them are at the limits of expected ranges.

Expected ranges for most parameters are shown in figure 11. These were determined from aquifer tests in the basin (Gannett and others, 2007) and modeling results from similar terrains in the upper Deschutes Basin (Gannett and Lite, 2004). The expected range of hydraulic conductivity values in the Cascade Range was derived from modeling work done by Manga (1996, 1997) and Ingebritsen and others (1992). Ranges of hydraulic-conductivity values for major rock types are also given in most groundwater texts such as Freeze and Cherry (1979) and Fetter (1980).

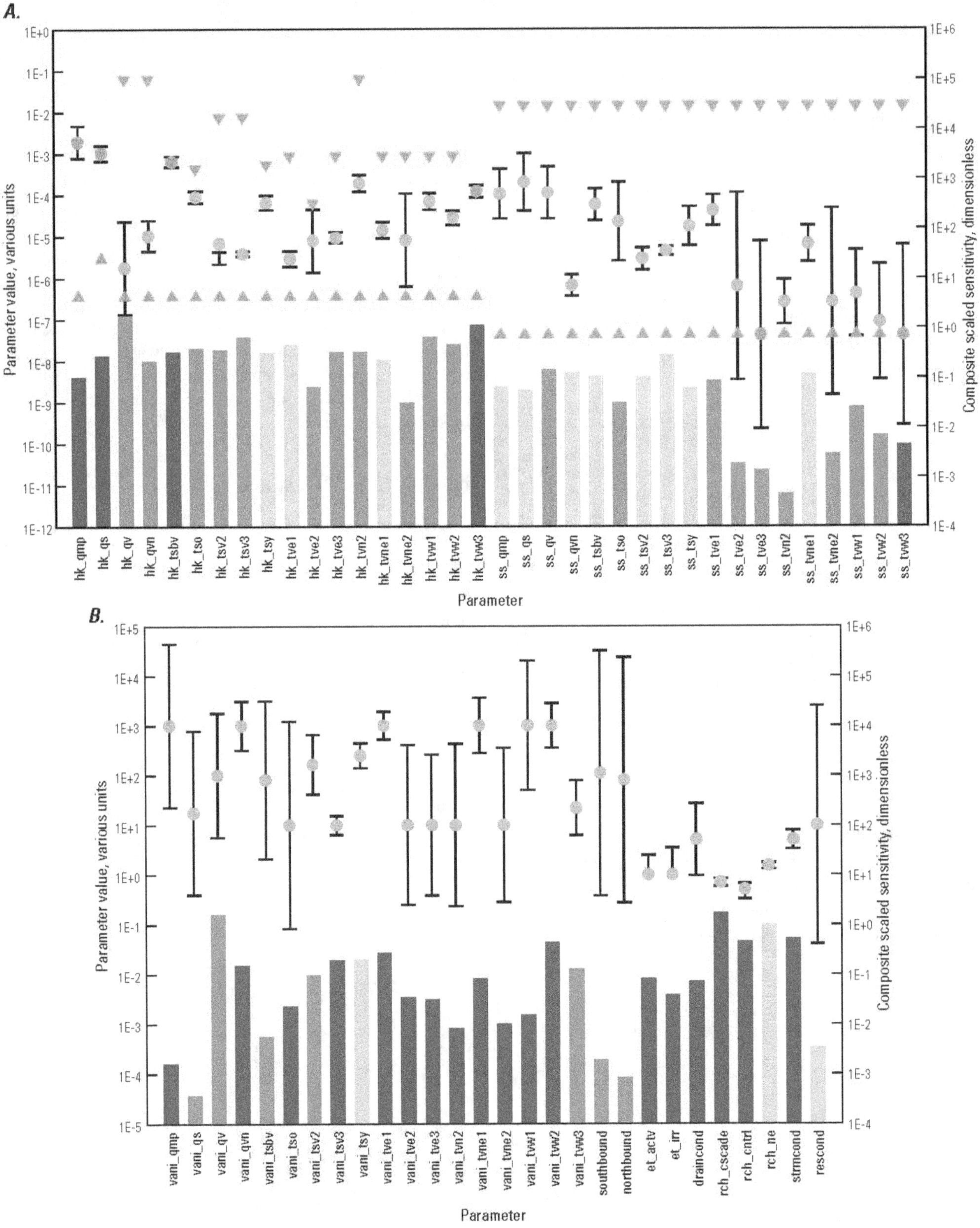

Figure 11. Final values (circles), expected ranges (triangles), and composite scaled sensitivities (bars) of parameters in the upper Klamath Basin, Oregon and California, regional groundwater flow model. Bars are shaded to indicate whether parameter values were fixed (light shading) estimated by PEST (medium shading), or estimated by PEST but at the limits of expected ranges (dark shading). Expected ranges are not shown on part *B* because there were no independent estimates.

Model Fit

Model fit describes the degree to which hydrologic conditions simulated by the model agree with observed conditions. Diagnostic and inferential statistics provide quantitative measures of model fit and are useful for comparing different models and quantifying model uncertainty. For most people, it is more intuitive to evaluate model fit using graphs and maps comparing simulated and measured heads and flows. Both approaches are discussed in this section.

Measures of Model Fit

The objective function $S(\underline{b})$ of equation 2 is a basic measure of model fit, but its usefulness for identifying model error and bias is limited. For these purposes, it is informative to evaluate the patterns of residuals (the differences between observed and simulated dependent variables). One desirable quality of residuals is that they be random and normally distributed. A useful tool for evaluating residuals is a graph of weighted residuals versus weighted simulated values (Hill, 1998; Hill and Tiedeman, 2007). In such graphs, it is desirable for residuals to be evenly distributed above and below zero, and for the entire dataset to show no slope or widening with respect to the x axis. Residuals plotted on figure 12 show no such trends as a group. Short linear trends within clusters in the dataset relate to the time series of individual well and streamflow datasets. These trends result from the amplitude of simulated fluctuations not matching exactly the observations. Overall, the graph shows a slight negative bias in heads, indicating that simulated heads tend to be too high more commonly than too low. The slight negative bias likely results from comparing head observations that are concentrated near the land surface and are clustered in lowland areas with simulated values that represent cell centers in relatively thick layers with upward vertical gradients.

A map of head residuals from the calibrated model (fig. 13) shows that the residuals are not spatially random but tend to cluster into areas of predominantly positive or negative residuals. Most head residuals are less than 10 ft, but larger residuals occur in the Butte Valley area and the Modoc Plateau.

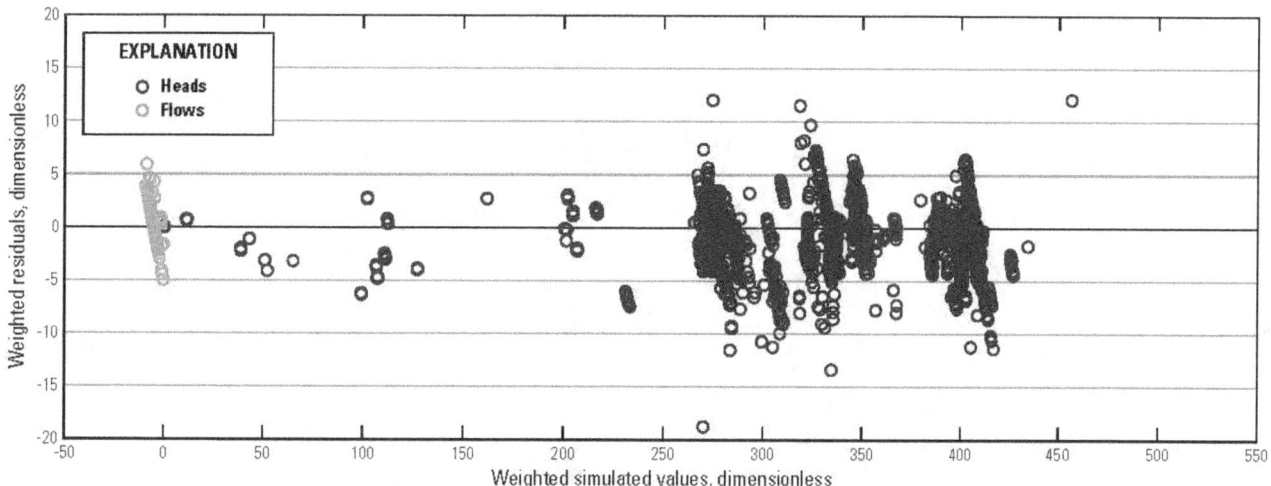

Figure 12. Weighted residuals plotted as a function of weighted simulated values.

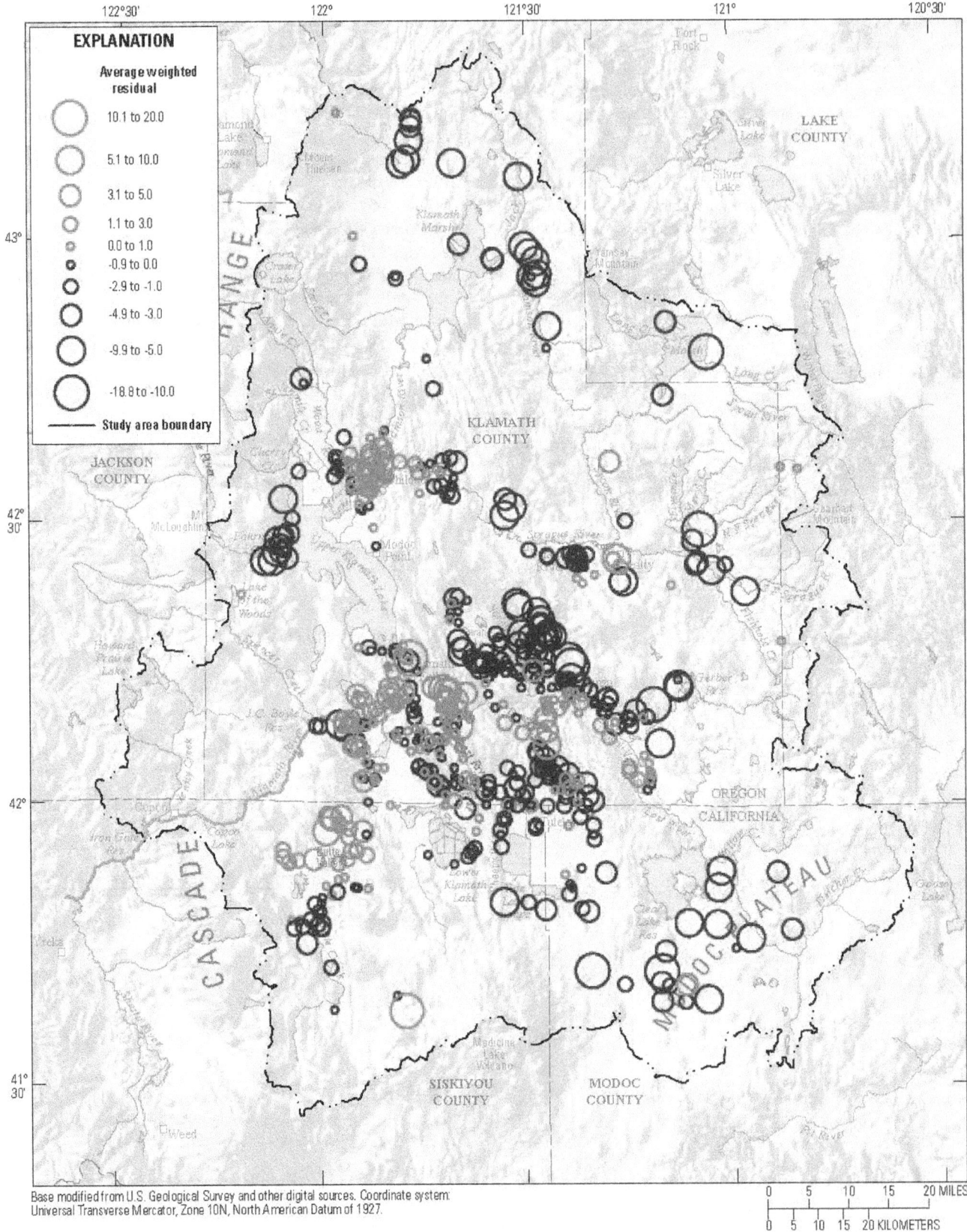

Figure 13.   Distribution of average weighted head residuals (dimensionless), upper Klamath Basin, Oregon and California. Positive values indicate simulated heads are lower than observed; negative values indicate simulated heads are higher than observed.

One measure of overall model fit is the calculated error variance, s^2, defined as

$$s^2 = \frac{S(\underline{b})}{(ND + PR - NP)}, \tag{6}$$

where

PR is the number of prior information values, and
NP is the number of parameters (Hill and Tiedeman, 2007, p. 95).

The square root of the calculated error variance, s, is known as the standard error of the regression. Smaller values of the calculated error variance and standard error of the regression indicate better model fit and are desirable. For the transient model calibration, $s^2 = 6.36$ and $s = 2.52$.

More intuitive measures of overall model fit are provided by fitted error statistics defined by Hill (1998). Fitted error statistics are derived by multiplying the standard error of the regression by the standard deviations or coefficients of variation used to define the weights for groups of observations, thus resulting in measures that have the same units as the observations. The weights for head observations in the transient model were based on a standard deviation of measurement error averaging about 12 ft. Given a standard error of 2.52, fitted standard deviation for heads is about 30 ft. This means that, in general, simulated heads match measured heads within a standard deviation of about 30 ft. For context, hydraulic heads span at least 3,700 ft in the upper Klamath Basin. Weights for flux observations were based on a coefficient of variation of about 0.2 (20 percent). Multiplying this by the standard error of the regression indicates that simulated fluxes of groundwater discharge to streams generally match measured fluxes with a fitted coefficient of variation of about 50 percent.

Graphical Comparison of Observed and Simulated Heads and Fluxes

Graphical depictions can provide an intuitive sense of model fit; they are especially useful when comparing measured and simulated time series such as fluctuations in water levels and groundwater discharge, and for evaluating model fit spatially.

Comparison of Observed and Simulated Heads

The geographic distribution of fit to hydraulic head is shown in figure 13. It can be seen that the fit is best, and residuals are smallest, in the interior parts of the basin where data are concentrated. The largest residuals are concentrated where wells are sparse, limiting the information on subsurface conditions, and where horizontal head gradients are steep, increasing sensitivity to spatial discretization effects.

Evaluating how well the model simulates temporal variations in hydraulic head can be evaluated using graphs comparing observed and simulated water-level time series. When comparing simulated and observed water-level time series it is important to be mindful that water levels in wells can be affected by external factors not simulated in the model such as stream-stage variations and pumping from nearby wells not included the model.

The hydrologic response to external stresses varies throughout the upper Klamath Basin due to differences in geology, hydrologic setting, and external stresses. Because of this, the discussion of observed and simulated water-level time series is organized by subbasins or geographic subareas in the basin. Figure 14 shows the locations of wells that are discussed in the following paragraphs.

Upper Williamson River Subbasin

Wells in the upper Williamson River subbasin typically show little or no seasonal water-level fluctuation. Of the several wells in the area with multiple water-level measurements, only one has data prior to 2000. All wells show monotonic water-level declines of 1.5 to 4 ft/yr since 2000 due to climate. The lack of seasonal water-level fluctuations may be due in part to the presence of a thick layer of pumice and other pyroclastic material from the eruption leading to the creation of the Crater Lake caldera that covers much of the principal recharge area. This highly porous material, which well data indicate can have an unsaturated thickness of more than 200 ft, slows the downward percolation of water and attenuates the seasonal variability of recharge. Where saturated, these clastic deposits have large storage coefficients, which also tend to dampen seasonal fluctuations.

Simulated water levels in the upper Williamson River subbasin tend to include seasonal variations not observed in the wells (fig. 15). This is likely due to the fact that the model does not simulate unsaturated zone processes that attenuate seasonal variations in recharge, and annual recharge pulses are applied directly to the saturated zone. The post-2000 water-level declines observed in most wells are accurately simulated (fig. 15). The fit to absolute heads is variable in the upper Williamson River. In the central and western parts of the area, simulated heads are within a few feet of measured heads. Near the northern and eastern margins of the area, however, weighted residuals range from 5 to 10 (dimensionless) equating to unweighted residual values of approximately 50 to 100 ft. The large residuals near the northeast margin may be an artifact of the no-flow boundary condition, and suggest there may be some inter-basin flow in that area.

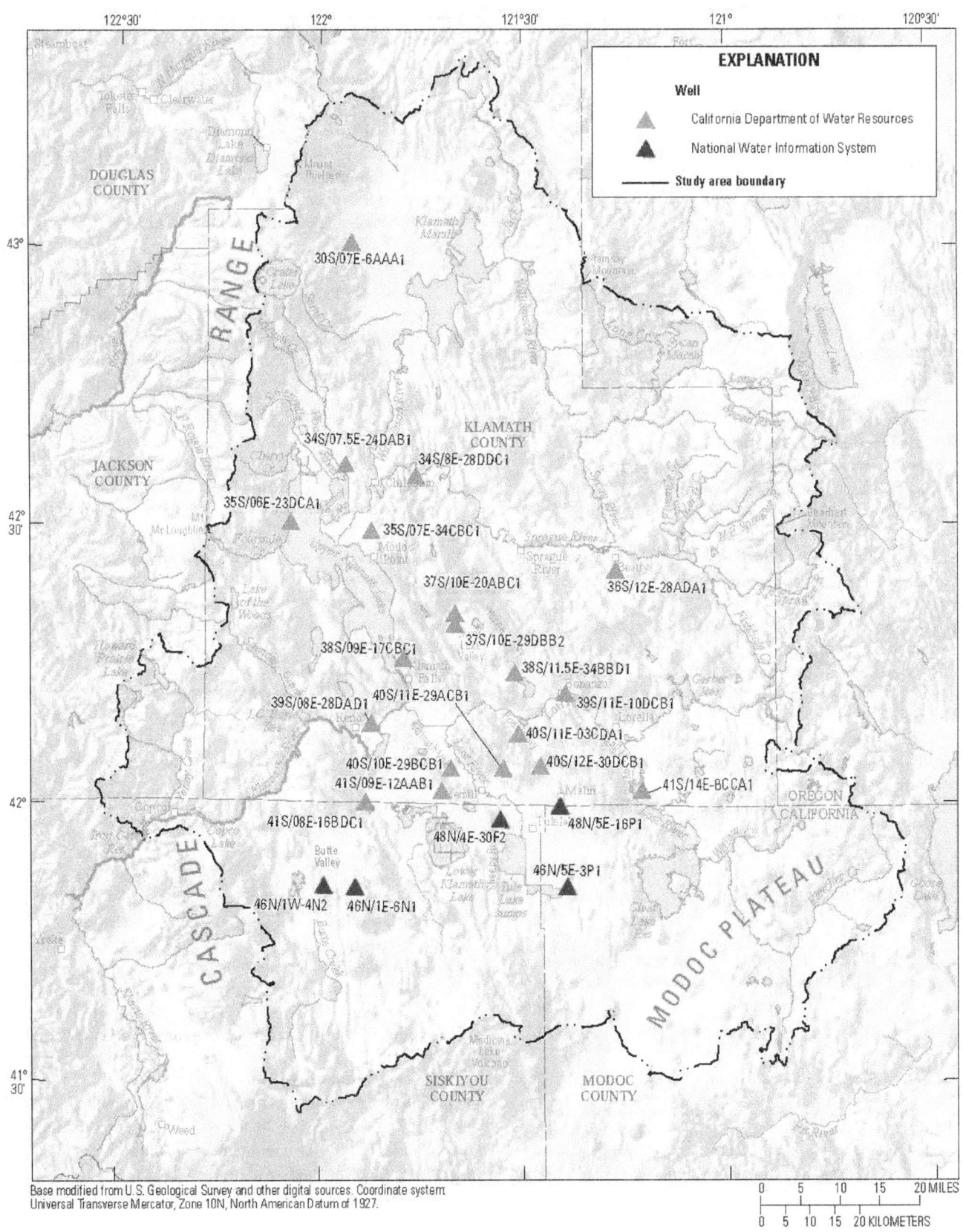

Figure 14. Locations of selected wells with water level time series in the upper Klamath Basin, Oregon and California.

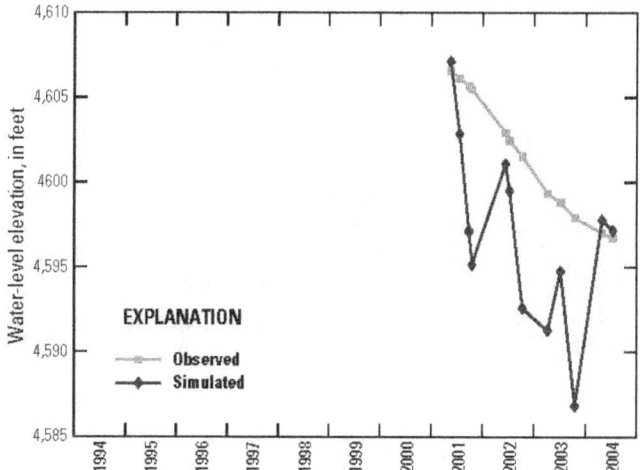

Figure 15. Observed and simulated water-level elevations in well 30S/7E-6AAA1 (OWRD Log ID KLAM 588) in the upper Williamson River subbasin, Oregon.

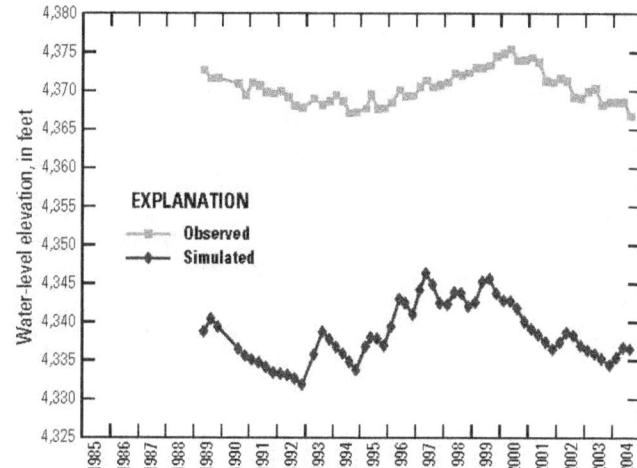

Figure 16. Observed and simulated water-level elevations in well 36S/12E-28ADA1 (OWRD Log ID KLAM 2096) in the Sprague River subbasin, Oregon.

Sprague River Subbasin

Seasonal variations observed in wells in the Sprague River subbasin are generally less than a few feet, unless they are influenced by nearby pumping. Observation wells with water-level measurement records going back to the late 1980s show climate-driven decadal-scale fluctuations of several feet. Most wells show slight climate-driven declines (approximately 1 ft/yr) since 2000.

The model fit in the uppermost Sprague River subbasin (including the area east of Bly and the Sycan Marsh) is variable. Average weighed residual values in the area range from –10 to 10 (fig. 13), corresponding to unweighted residuals of roughly –100 to 100 ft. The fit is better near Beatty where residuals are generally less than 15 ft. Decadal scale climate-driven head fluctuations near Whiskey Creek are simulated by the model, but are slightly larger than observed (fig. 16) and the simulated peak precedes the observed peak by several months. Downstream from Beatty, model fit is variable, with residuals ranging from approximately –60 ft to 8 ft. Simulated water levels generally show seasonal fluctuations of 1 to 3 ft, while observed water levels show smaller or no variations (fig. 17). This slight discrepancy may stem from the fact that the attenuating effects of large near-surface storage coefficients are not represented in the coarse vertical discretization in the regional model.

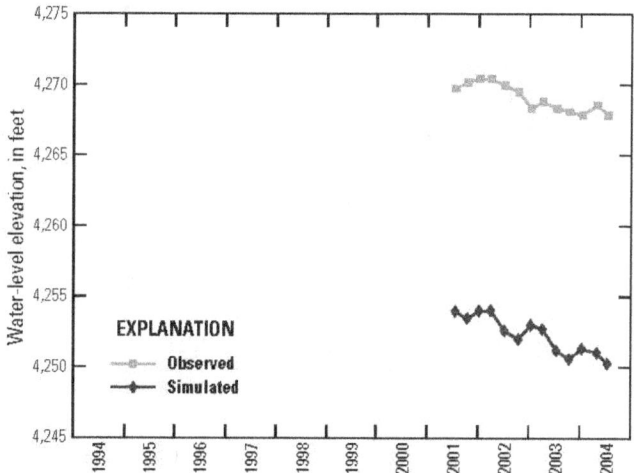

Figure 17. Observed and simulated water-level elevations in well 34S/8E-28DDC1 (OWRD Log ID KLAM 1055) in the Sprague River subbasin, Oregon.

Wood River Valley

The period of record for most wells with water-level time series in the Wood River subbasin extends back only to about 2000. Calibration data extend back to 1989 for only one well. Observed water levels in wells in the Wood River Valley commonly show seasonal fluctuations of a few feet in response either to seasonal variations in the stage of Upper Klamath Lake or to recharge, depending on location. The single well with data back to the late 1980s (near Modoc Point) (fig. 18) shows a climate-driven decadal fluctuation of about 2 ft in addition to the seasonal fluctuation of 2 to 3 ft. The model simulates a decadal fluctuation of about 4 feet and seasonal fluctuations of about 1 ft (fig. 18).

Simulated heads on the west side of the Wood River Valley near Rocky Point are 50 to 70 ft higher than observed, likely as a result of coarse vertical discretization in an area of strong upward gradients. Simulated seasonal fluctuations are larger than observed fluctuations, with a possible lag in timing in some instances by about 3 months (one stress period) (fig. 19). Simulated water levels are generally within about 5-10 ft of observed levels on the east side of the Wood River Valley. Climate-driven water level declines east of the lake are captured by the model but are slightly steeper than observed (fig. 20).

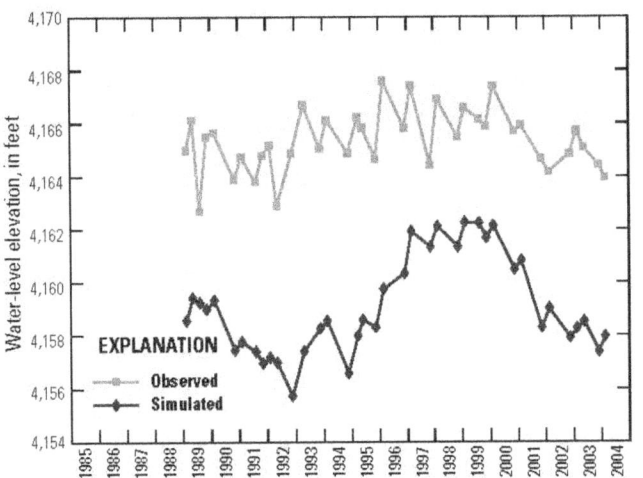

Figure 18. Observed and simulated water-level elevations in well 35S/7E-34CBC1 (OWRD Log ID KLAM 1362) in the Wood River subbasin, Oregon.

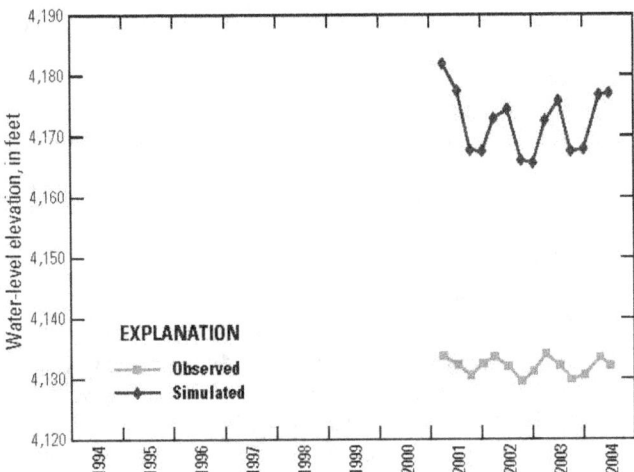

Figure 19. Observed and simulated water-level elevations in well 35S/6E-23DCA1 (OWRD Log ID KLAM 1125) in the Wood River subbasin, Oregon.

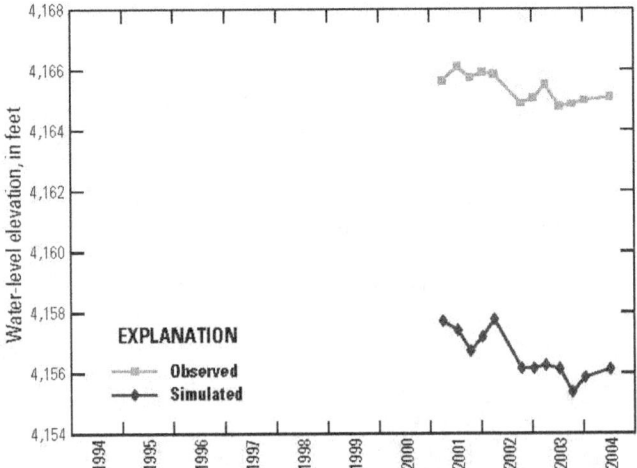

Figure 20. Observed and simulated water-level elevations in well 34S/7.5E-24DAB1 (OWRD Log ID KLAM 1007) in the Wood River subbasin, Oregon.

Swan Lake Valley

Several wells are used for calibration in the Swan Lake Valley area, some with periods of record extending back to the late 1980s. Water-level records show decadal-scale water level fluctuations of about 5 ft that appear to be mostly climate driven. In addition, there are seasonal fluctuations on the order of 1 to 5 ft, depending on location.

Simulated water levels are within 15 ft of observed levels in most wells in the Swan Lake Valley, with one well near the western margin in which simulated heads are about 70 ft higher than observed. The model simulated the decadal-scale water-level fluctuation reasonably well (fig. 21). Seasonal fluctuations are simulated (fig. 22), but amplitudes did not always match exactly, probably due to differences between simulated and actual pumping centers.

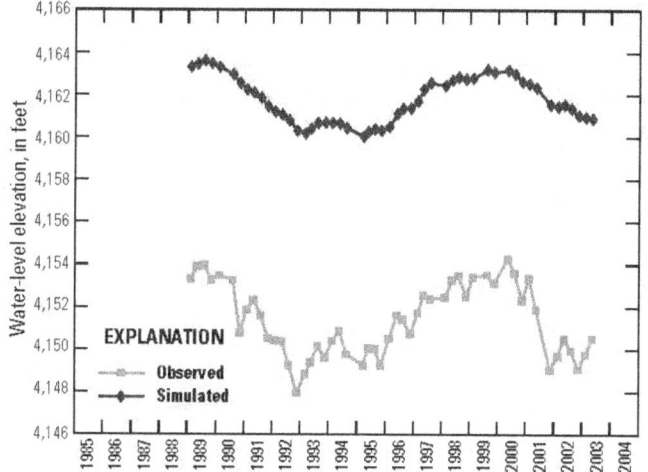

Figure 21. Observed and simulated water-level elevations in well 37S/10E-29DBB2 (OWRD Log ID KLAM 2288) in the Swan Lake Valley area, Oregon.

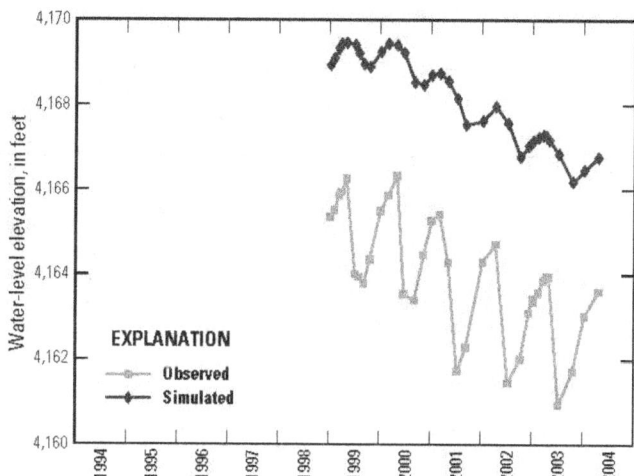

Figure 22. Observed and simulated water-level elevations in well 37S/10E-20ABC1 (OWRD Log ID KLAM 2277) in the Swan Lake Valley area, Oregon.

Upper Lost River Subbasin

The upper Lost River subbasin contains a rich set of water-level data collected by OWRD, with some records extending back to the late 1980s (Grondin, 2004). Water levels in wells exhibit decadal-scale and seasonal fluctuations of up to approximately 5 ft. Local deviations from climate-driven and seasonal fluctuations are caused by changes in pumping patterns. The magnitude of fluctuations varies geographically throughout the upper Lost River subbasin.

The absolute difference between observed and simulated heads varies geographically in the upper Lost River subbasin. Simulated heads are usually within about 15 to 20 ft of observed heads throughout the area. There are a few wells with larger residuals in all areas. Residuals are consistently large, ranging from 40 to 70 ft, in the Yonna Valley.

Long-term (decadal-scale) water-level fluctuations are generally underestimated by the model in the upper Lost River subbasin (fig. 23). Seasonal and recent interannual trends are simulated accurately in some wells (fig. 24), but interannual trends are not captured in all wells, probably due to head-dependent flux boundaries (fig. 25). Water-level perturbations caused by year-to-year variations in pumping are simulated by the model where the pumping data exist (fig. 26).

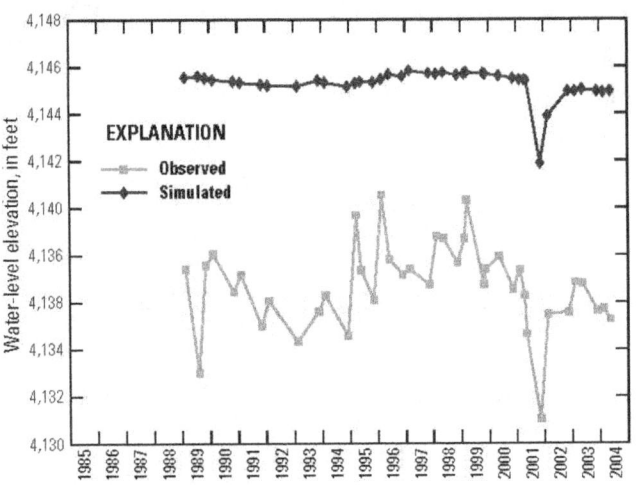

Figure 23. Observed and simulated water-level elevations in well 41S/14E-8CCA1 (OWRD Log ID KLAM 15130) in the upper Lost River subbasin, Oregon.

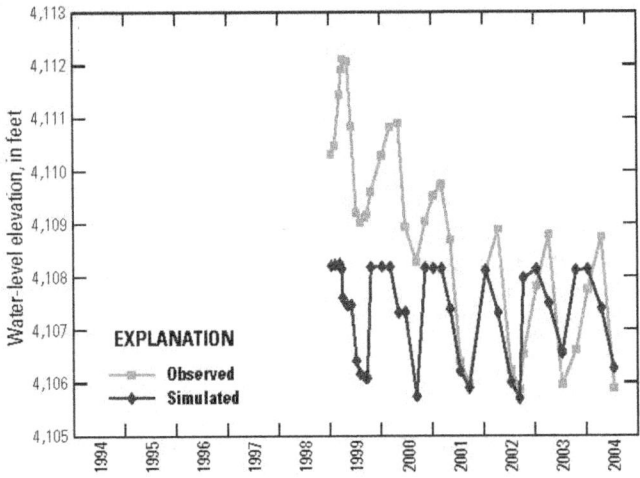

Figure 25. Observed and simulated water-level elevations in well 39S/11E-10DCB1 (OWRD Log ID KLAM 51922) in the upper Lost River subbasin, Oregon.

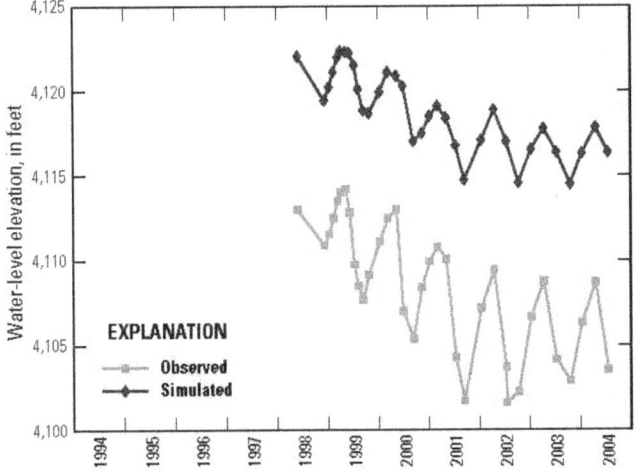

Figure 24. Observed and simulated water-level elevations in well 38S/11.5E-34BBD1 (OWRD Log ID KLAM 11139) in the upper Lost River subbasin, Oregon.

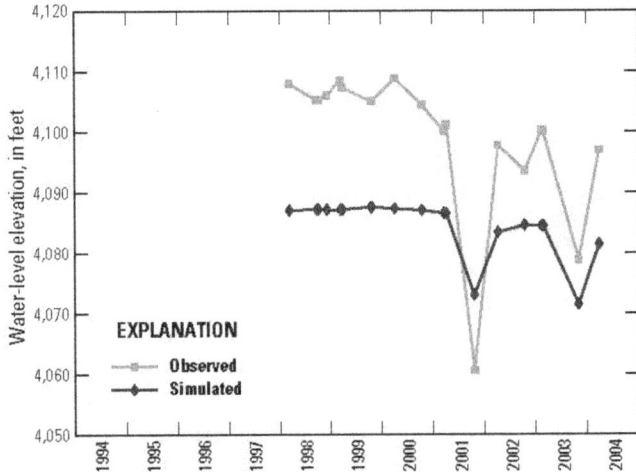

Figure 26. Observed and simulated water-level elevations in well 40S/11E-3CDA1 (OWRD Log ID KLAM 50632) in the upper Lost River subbasin, Oregon.

Klamath Falls/Klamath Valley Areas

Although long-term water-level records suitable for model calibration in the Klamath Falls/Klamath Valley area are lacking, several wells have records beginning in the late 1990s. The lack of long-term records makes climate trends difficult to discern. Water-level observations show seasonal variations ranging up to 15 ft. Because of the variable influence of canal leakage, irrigation, and pumping, water-level fluctuation patterns vary geographically and year to year.

In general, simulated heads match observed heads in the area within about 15 ft; one well near Merrill has average residuals of about 25 ft. Because stresses in the model are only varied quarterly, and because of the general lack of accurate information on the year-to-year spatial distribution of pumping and canal leakage, the spatial and temporal variability in observed seasonal water-level fluctuations is not matched in most wells. Seasonal fluctuations are simulated reasonably well north of Klamath Falls (fig. 27), where canal influences are minimal and pumping patterns are stable. Where accurate pumping information is available (as is the case in areas influenced by the pilot water bank pumping) the model simulates the response quite accurately (fig. 28).

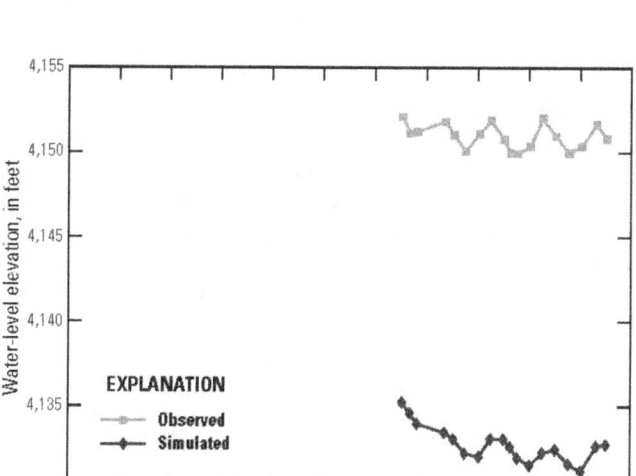

Figure 27. Observed and simulated water-level elevations in well 38S/9E-17CBC1 (OWRD Log ID KLAM 11656) in the Klamath Valley area, Oregon.

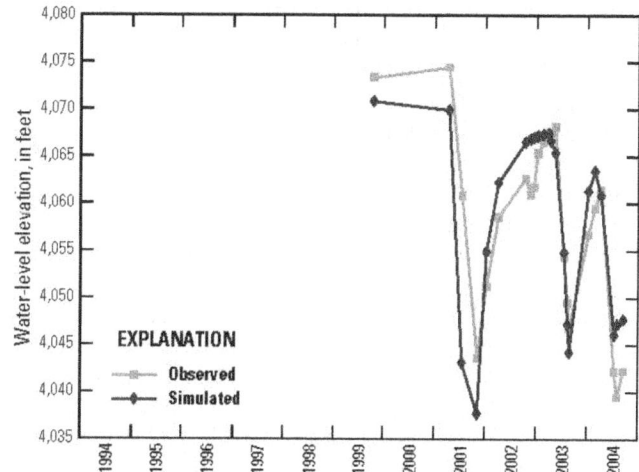

Figure 28. Observed and simulated water-level elevations in well 40S/10E-29BCB1 (OWRD Log ID KLAM 10518) in the Klamath Valley area, Oregon.

Tule Lake Subbasin

Like the Klamath Valley to the north, the Tule Lake subbasin is hydrologically complex. Water levels in the area reflect a wide range of external stresses including canal leakage, irrigation, groundwater pumping, and climate influences. As a consequence, water-level fluctuations vary geographically and from year to year. The dominant change in stress in the area is the large increase in groundwater pumping starting in 2001 in response to surface-water shortages. Prior to 2001, the area was characterized by relatively stable water levels, with modest seasonal and interannual variations. The pumping increase resulted in seasonal water-level declines (due to drawdown) of tens of feet, and interannual declines ranging from a few to 10 ft. Post 2000 pumping patterns vary from year to year because pumping rates and locations of pumped wells varied each year. Model calibration in the Tule Lake subbasin was aided by the large amount of pumping and water-level data collected by OWRD, USGS, CDWR, and Reclamation since 2000.

In the southern Tule Lake subbasin, including the Copic Bay area, simulated water levels are generally within about 20 ft of observed levels, with a few wells showing residuals of 40 to 50 ft. The large residuals generally occur south of the subbasin. Post 2000 pumping signals in the southeast part of the Tule Lake subbasin are simulated by the model well (fig. 29).

Deep wells near the State line exhibited a strong response to post 2000 pumping due to the large-capacity Tulelake Irrigation District (TID) wells arrayed along the border. The model simulates the observed acute drawdown due to seasonal pumping with reasonable accuracy (figs. 30 and 31). Because simulated heads are averaged across 2,500 by 2,500 ft cells, drawdown in, or very close to, actively pumped wells cannot be accurately simulated by the model. Water levels strongly affected by pumping were removed from the calibration dataset.

In some cases, the timing of observed drawdown does not coincide with drawdown simulated by the model. This is because pumping volumes were often provided as quarterly or yearly totals and the exact timing and rate of pumping were unknown. In addition, pumping (and the resulting measured drawdown) may have commenced part way through the stress period.

Along the northern margin of the Tule Lake subbasin (north of Merrill and Malin), recent seasonal head elevations and fluctuations are reasonably well simulated (fig. 32). The step-like decline in water levels in parts of the Tule Lake subbasin observed after 2000 is reasonably well simulated in some wells (fig. 29), but not fully captured in other locations (fig. 33). This is likely due to the limited information on the exact locations and rates of pumping, as well as a lack of information on subsurface geology.

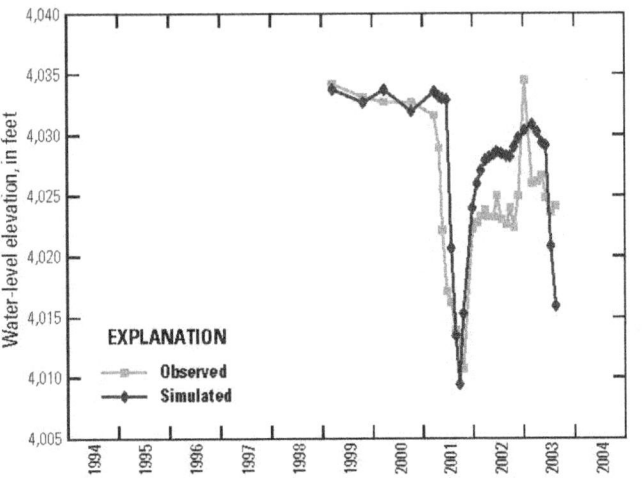

Figure 29.   Observed and simulated water-level elevations in well 46N/05E-3P1 (CDWR Well No. 46N05E03P001M) in the southern Tule Lake subbasin, California.

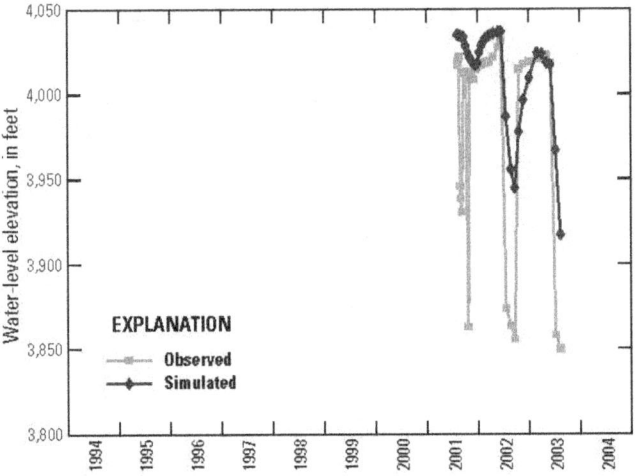

Figure 30.   Observed and simulated water-level elevations in well 48N/5E-16P1 (CDWR Well No. 48N05E16P001M) in the Tule Lake subbasin, California.

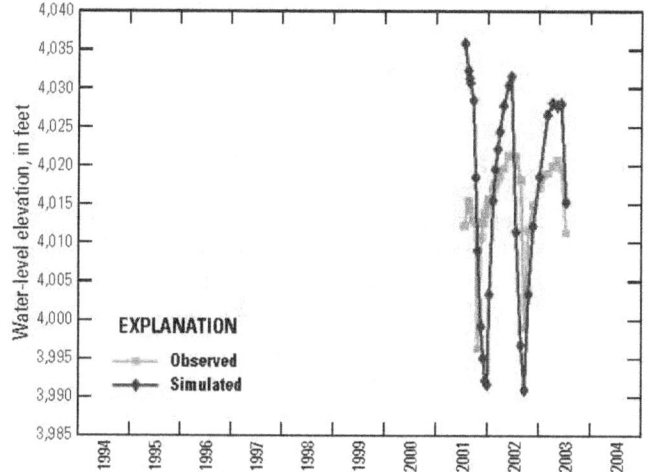

Figure 31. Observed and simulated water-level elevations in well 48N/04E-30F2 (CDWR Well No. 48N04E30F002M) in the Tule Lake subbasin, California.

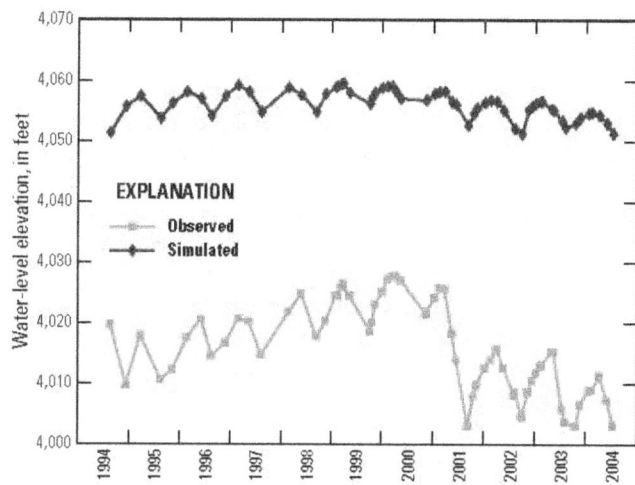

Figure 33. Observed and simulated water-level elevations in well 40S/12E-30DCB1 (OWRD Log ID KLAM 14829) in the Tule Lake subbasin, Oregon.

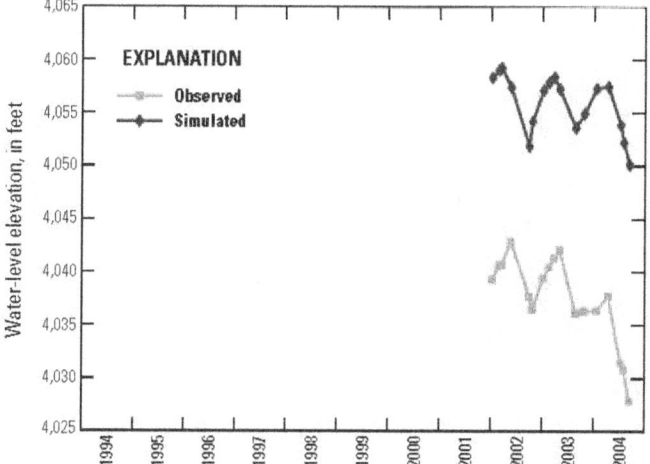

Figure 32. Observed and simulated water-level elevations in well 40S/11E-29ACB1 (OWRD Log ID KLAM 14764) in the northern Tule Lake subbasin, Oregon.

Lower Klamath Lake Subbasin

Monitored wells in the Lower Klamath Lake subbasin fluctuate in response to a variety of influences including the stage in managed wetlands, pumping, and climate. Climate signals, although present, are small and difficult to discern due to the short period of record for most wells. Seasonal fluctuations are spatially variable because of the geographic diversity of pumping and wetland management stresses.

Simulated heads are generally within 5 to 10 ft of observed heads in the Lower Klamath Lake subbasin. The fit between simulated and observed seasonal fluctuations is spatially variable because of the lack of accurate information on rates and locations of pumping. Seasonal head fluctuations are simulated with reasonable accuracy in some areas, such as the western and northern parts of the subbasin (figs. 34 and 35), but less so in the eastern parts of the basin, near the south end of the Klamath Hills (fig. 36). The well represented in figure 36 is the only one in the calibration dataset in the Lower Klamath Lake subbasin with sufficient record to capture decadal climate effects, but the increased seasonal fluctuations during the dry periods in the early 1990s and post 2000 time period suggest that the decadal signal in this well may be influenced by changes in pumping as well as drought.

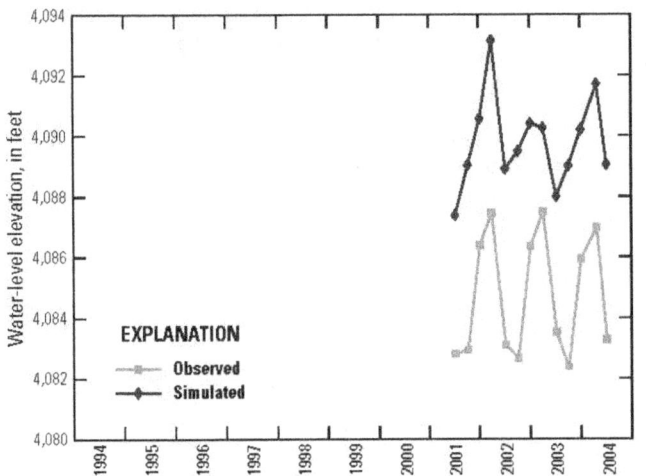

Figure 34. Observed and simulated water-level elevations in well 41S/8E-16BDC1 (OWRD Log ID KLAM 50228) in the Lower Klamath Lake subbasin, Oregon.

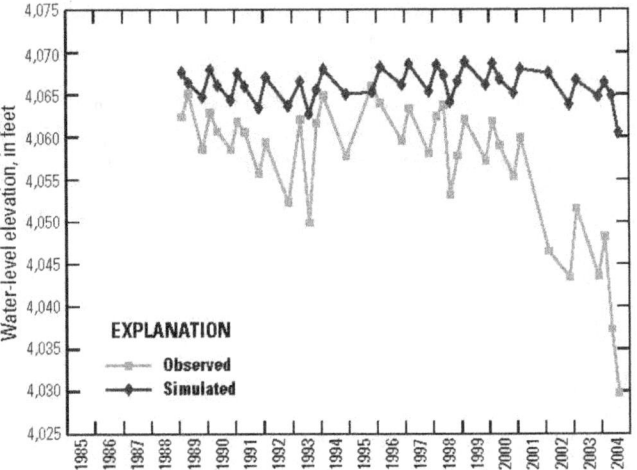

Figure 36. Observed and simulated water-level elevations in well 41S/9E-12AAB1 (OWRD Log ID KLAM 14914) in the Lower Klamath Lake subbasin, Oregon.

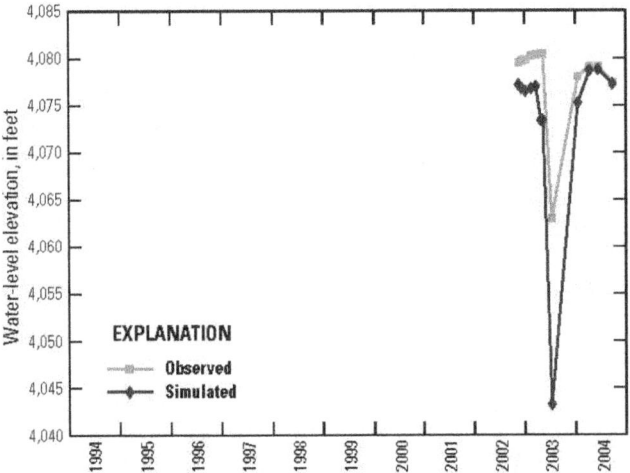

Figure 35. Observed and simulated water-level elevations in well 39S/8E-28DAD1 (OWRD Log ID KLAM 53320) in the Lower Klamath Lake subbasin, Oregon.

Butte Valley Area

A rich data set has been collected by CDWR in the Butte Valley area extending to the late 1980s. Most wells show a decadal climate signal of 5 to 10 ft as well as seasonal pumping signals generally ranging from 0 to 10 ft. Simulated heads are generally within about 10 to 30 ft of observed heads in most of the Butte Valley area. Several wells have residuals in the 40 to 50 ft range in the northern part. Simulated heads are systematically low throughout the area. The model does an excellent job of simulating the observed decadal climate signal

(figs. 37 and 38). The match between simulated and observed seasonal fluctuations is variable. Where differences occur, simulated fluctuations tend to be smaller than observed. This is likely due to the averaging of pumping effects over model grid cells as well as the way pumping stresses were distributed in the model. Because data were not available to assign pumping rates to individual wells, pumping stresses were assigned to the centers of all groundwater-irrigated fields identified by CDWR in their 2000 land use survey.

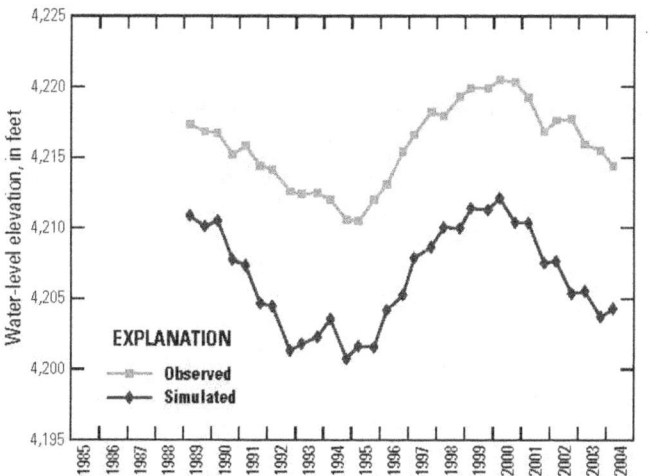

Figure 37. Observed and simulated water-level elevations in well 46N/1W-4N2 (CDWR Well No. 46N01W04N002M) in the Butte Valley area, California.

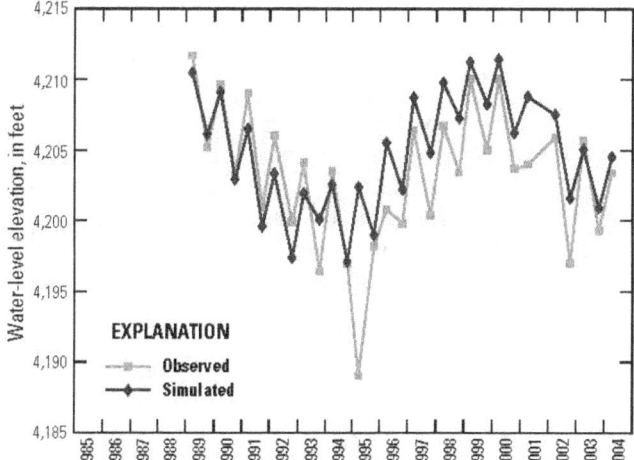

Figure 38. Observed and simulated water-level elevations in well 46N/1E-6N1 (CDWR Well No. 46N01E06N001M) in the Butte Valley area, California.

Comparison of Observed and Simulated Groundwater Discharge to Streams

The model is calibrated using groundwater-discharge data in addition to hydraulic-head data. Groundwater-discharge measurements or estimates are available for numerous streams, stream reaches, or groups of streams in the basin. Estimates of long-term average groundwater discharge were available for 52 locations throughout the basin (table 3). Measurements or estimates of the temporal variations in groundwater discharge necessary for transient calibration are less common. Data suitable for model calibration were available for only 10 areas in the upper Klamath Basin.

Simulated discharge averaged over the transient calibration period is compared to the 52 reaches with long-term average discharge estimates in table 3. Observed values and confidence intervals derived from table 6 in Gannett and others (2007) are shown with the simulated equivalents in figure 39. Figure 39 shows that the majority of the simulated values are within or close to the expected ranges, and that most areas with large groundwater discharge are well simulated.

The model's ability to simulate variations in groundwater discharge in response to external stresses can be evaluated by graphs comparing simulated and observed groundwater discharge to streams. Time series of groundwater discharge to streams suitable for model calibration were available for ten stream reaches.

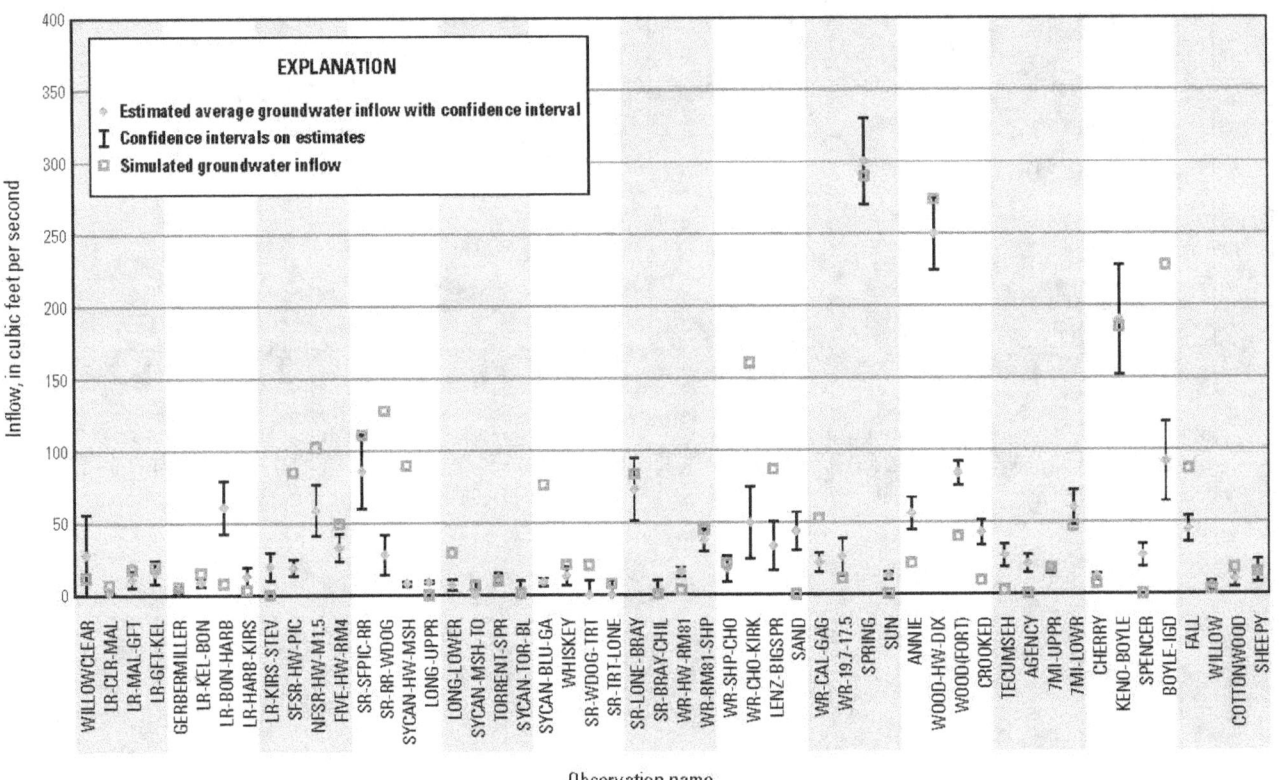

Figure 39.   Observed and simulated long-term average groundwater discharge to selected stream reaches in the upper Klamath Basin, Oregon and California. Observation names relate to data in table 3.

Observed and simulated groundwater discharge to an aggregate of reaches in the upper Sprague River system are shown in figure 40. This reach group includes streams with substantial groundwater discharge in the Sprague River drainage above the gage at Beatty. Simulated groundwater discharge is compared with observed groundwater discharge based on September mean streamflows at the Beatty gage, which are considered a reasonable proxy for base flow. The September mean flows of the Sprague River reflect a slight drying trend since the early 1970s with superimposed decadal drought cycles. Simulated values are slightly larger than observed values, but the slight long-term downward trend, decadal cycles, and interannual variations are well matched.

Simulated groundwater discharge to the upper Williamson River (above the gage near Sheep Creek) is shown in figure 41. This reach includes the main-stem Williamson River from the headwaters to the gage. As with the upper Sprague River, the observed groundwater discharge is based on the September mean streamflows and compared with simulated values at the appropriate time step. Simulated groundwater discharge to the upper Williamson River is slightly lower than estimated, but the long-term trend, decadal cycles, and interannual variations are reasonably well captured.

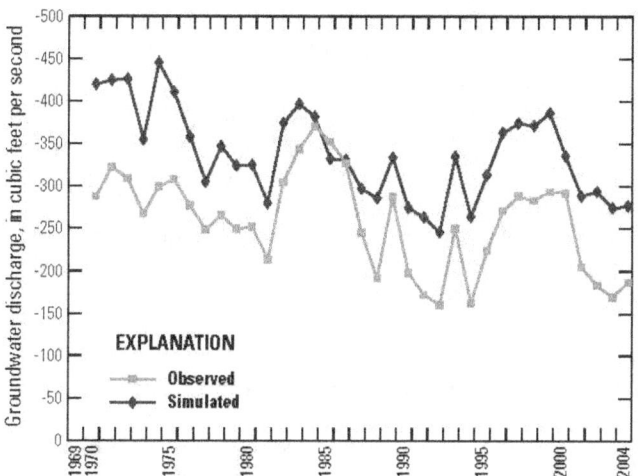

Figure 40.   Observed and simulated groundwater discharge to the upper Sprague River, Oregon. Observed groundwater discharge based on September mean flow.

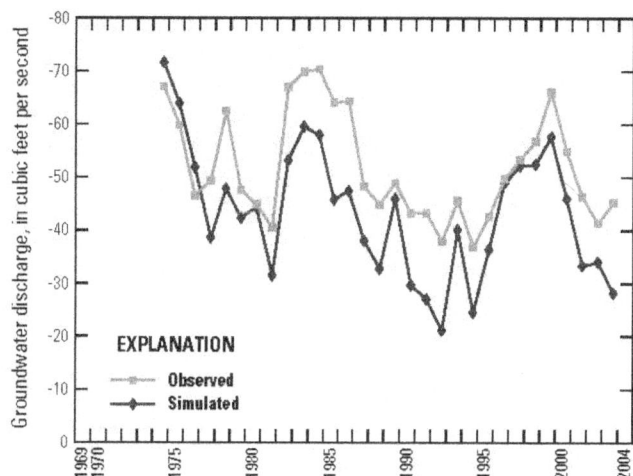

Figure 41.   Observed and simulated groundwater discharge to the upper Williamson River, Oregon. Observed groundwater discharge based on September mean flow.

The Williamson River between Kirk and the Sprague River is a major groundwater-discharge area. There is considerable groundwater discharge to the Williamson River and tributaries, including Spring Creek, in this reach. This discharge is quantified using data from gaging stations on the Williamson River near Kirk and below the Sprague River, and on the Sprague River near Chiloquin. Although this is a major center of groundwater discharge in the basin and the general magnitude of groundwater inflow is well quantified, monthly estimates of groundwater discharge are uncertain. Part of the uncertainty is attributable to the three gaging stations, each with its own measurement error, used to estimate inflow. This uncertainty is largest during high-flow periods in winter and spring when measurement error is large compared to groundwater inflow. Another source of uncertainty is ungaged diversions, primarily the diversion for the Modoc Irrigation District, which must be estimated from historic measurements. Error due to ungaged diversions is largest during the summer irrigation season. To reduce the effects of measurement error, simulated groundwater inflow to the lower Williamson River is compared to an observation based on a 5-month running average of measured inflow for model calibration. A graph of observed and simulated inflow (fig. 42) shows that the long-term and decadal trends apparent in the observed values, as well as the overall volume of groundwater discharge, are

simulated reasonably well. For example, the general decline in groundwater discharge between the early 1980s and mid 1990s is captured, as is the increase in discharge of the late 1990s (after the gap in the data due to non operation of the gage at the outlet of Klamath Marsh near Kirk [USGS gage 11403500]). Seasonal variations are less consistently simulated, possibly due to a combination of measurement error and model error.

Observations of groundwater discharge to the Wood River headwaters are based on instantaneous measurements of the discharge of the Wood River made just below the headwater springs (at Dixon Road). Flow at this site is entirely groundwater and is unaffected by diversions or other inflows. Measurements indicate that discharge to the headwater springs of the Wood River has seasonal variations of a few tens of cubic feet per second and interannual flow variations of between 50 and 100 ft³/s. In general, the largest changes observed are increases in flow after wet winters; such increases are usually followed by more gentle recession curves. Simulated and observed groundwater discharge to the Wood River headwaters are shown in figure 43. The general magnitude of discharge and the timing of variations are well simulated, but the simulated magnitude of temporal variations tends to be less than observed.

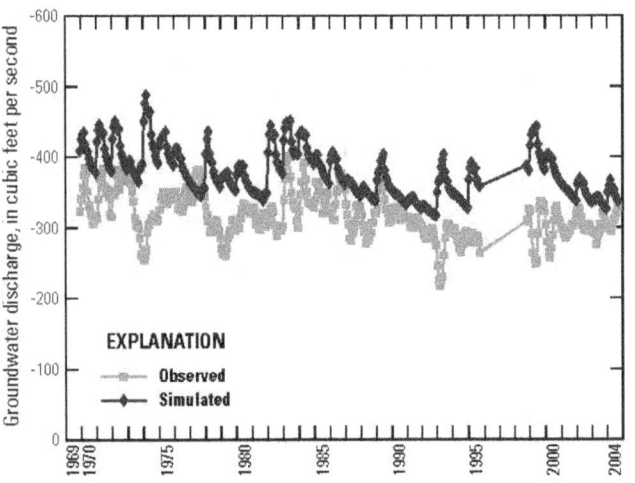

Figure 42. Observed and simulated groundwater discharge to the lower Williamson River, Oregon. Observed groundwater discharge calculated from stream-gage data.

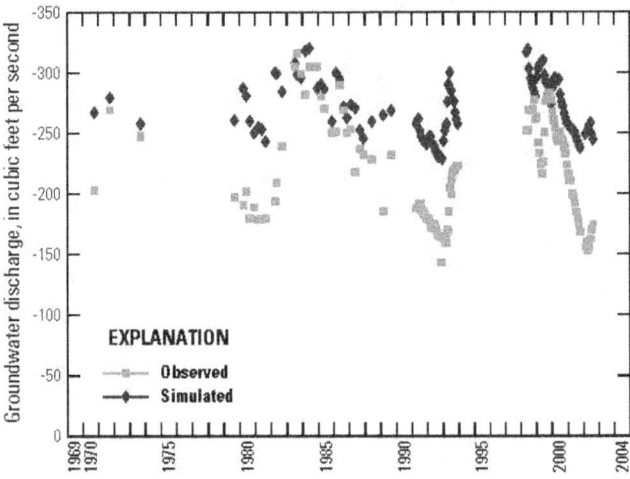

Figure 43. Observed and simulated groundwater discharge to the headwaters of the Wood River, Oregon.

Short periods of streamflow record were available for Cherry Creek and Sevenmile Creek, tributaries to Upper Klamath Lake that flow from the Cascade Range. Because these streams include both groundwater and runoff components, September mean flows were used as a proxy for groundwater discharge. Figures 44 and 45 show observed and simulated September mean groundwater discharge to Sevenmile and Cherry Creeks, respectively. The average rates of groundwater discharge to both of these streams are reasonable; however, the model simulates larger interannual variations than observed. For Cherry Creek, simulated September groundwater discharge is zero some years. Overestimation by the model of temporal variations in groundwater discharge in some areas results from the lack of unsaturated-zone processes in the model. Water-level data in the basin suggest that thick unsaturated zones tend to attenuate interannual variations. Because this is not accounted for in the present model, simulated temporal variations in head and discharge tend to be larger than observed in some areas.

A short period of streamflow record is also available for Spencer Creek, tributary to the Klamath River in John C. Boyle Reservoir. As with Sevenmile and Cherry Creeks, observed groundwater discharge is based on September mean flow. Simulated groundwater discharge to Spencer Creek is about three times larger than observed (fig. 46); however, temporal variations are simulated more accurately. The large simulated discharge to Spencer Creek is not unexpected as it is the only stream simulated in the area to which groundwater can discharge. Other streams draining the general area, such as the upper reaches of the Jenny Creek, are not simulated. Therefore, any groundwater discharge to these streams must be accommodated in the model by Spencer Creek.

Gains to the Klamath River between the gages at Keno and below John C. Boyle Reservoir are estimated by subtracting monthly mean flows at the bounding gages and correcting for changes in reservoir storage. August mean gains in flow are considered a reasonable proxy for groundwater discharge to this reach. As calculated, however, these observed gains include flows from ungaged tributaries such as Spencer Creek, so observed inflows to the Klamath River may be slightly overestimated. August mean flows of Spencer Creek, the largest tributary to this reach, averaged about 22 ft³/s from 1993 to 1997. A comparison of observed and simulated groundwater discharge to the Klamath River in this reach (fig. 47) shows that simulated values are slightly lower than the estimated values. Simulated discharge reflects year-to-year variations in recharge due to climate, but larger-amplitude decadal variations due to drought cycles are not fully captured. The extent to which the decadal variations in the estimated inflows result from the effects of ungaged tributaries is not known.

Simulated heads in the area of Sand Creek were below the stream, so simulated groundwater discharge is zero. Simulated groundwater discharge to Bonanza Springs is only about 10 percent of the observed mean discharge.

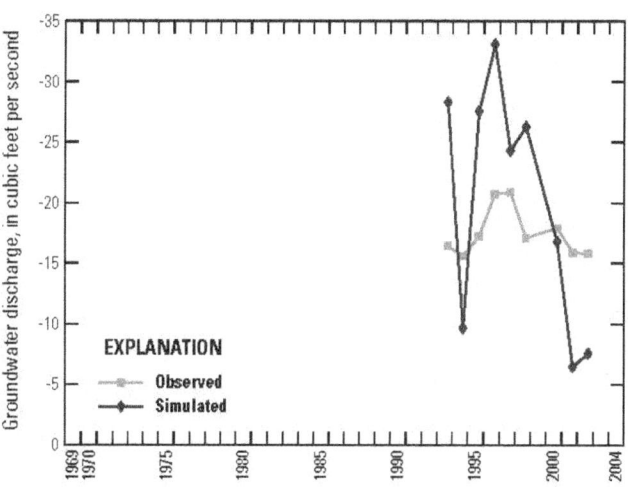

Figure 44. Observed and simulated groundwater discharge to upper Sevenmile Creek, Oregon. Observed groundwater discharge based on September mean flow.

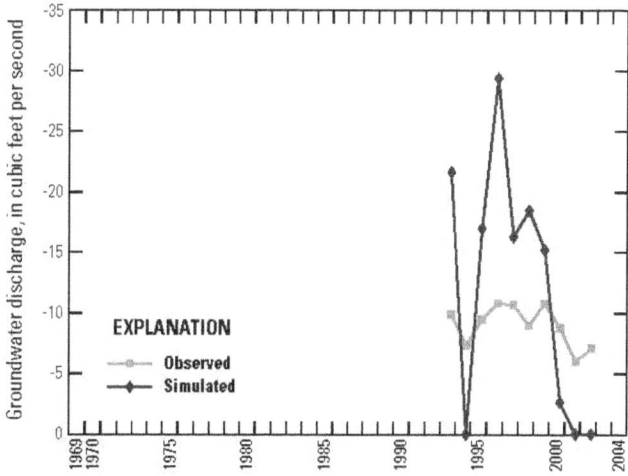

Figure 45. Observed and simulated groundwater discharge to upper Cherry Creek, Oregon. Observed groundwater discharge based on September mean flow.

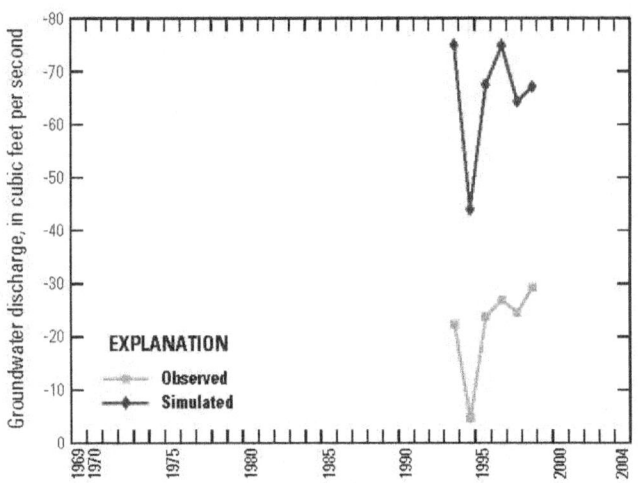

Figure 46. Observed and simulated groundwater discharge to Spencer Creek, Oregon. Observed groundwater discharged based on September mean flow.

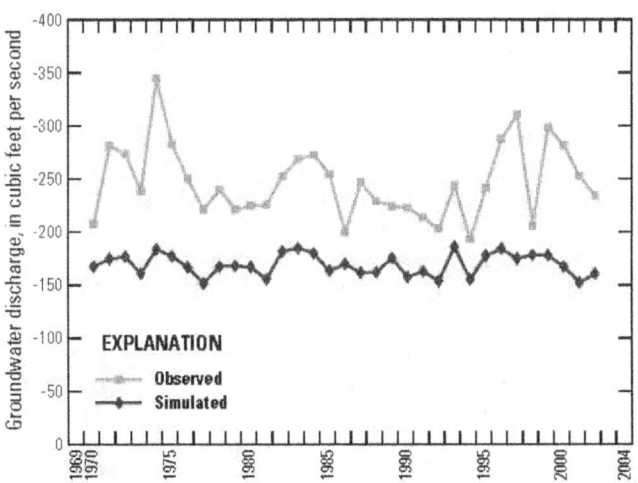

Figure 47. Observed and simulated groundwater discharge to the Klamath River between the gaging stations at Keno and below John C. Boyle Dam, Oregon. Observed groundwater discharge based on gains in streamflow calculated from stream gage and reservoir stage data.

Example Simulations

Example simulations are presented in this section to demonstrate the basic capabilities of the model and to show how effects of groundwater pumping vary depending on location. Simulations are presented for three areas selected to show a range of responses. All simulations in this section involve pumping a single well at 10 ft³/s continuously during the fourth quarter of the water year (July through September). This is intended to approximate the pumping of a large-capacity well during much of an irrigation season. A mean annual pumping rate of 2.5 ft³/s results from pumping 10 ft³/s for 3 months of the year.

The example simulations are run in a dynamic steady state mode, meaning that they are transient simulations in which all external stresses other than the single well being evaluated, such as recharge and background pumping, vary quarterly but not interannually. Quarterly external stresses are based on values from the 1980 water year, which most closely represents average conditions during the 1970 to 2004 period. Background pumping in the dynamic steady-state simulations is based on 2000 pumping rates. To produce starting heads for the dynamic steady state simulations, multiple 50-year simulations were run using the final heads from the preceding simulation as starting heads. This process was repeated until background simulations had no long term trends. The example simulations were all run for 50 years, with pumping starting in the third year.

For each of the simulations, the main processes discussed are drawdown (specifically, the change in groundwater storage) and the effects of pumping on hydrologic boundaries such as streams, springs, lakes, and agricultural drains. At the onset of pumping, most water pumped from wells comes from storage. The removal of water from storage results in a lowering of the water table in a cone-shaped area around the well known as the cone of depression. As the cone of depression expands, groundwater flow is redirected toward the well, affecting flow to and from hydrologic boundaries such as streams and springs. The cone of depression stabilizes when the changes in flow to and from hydrologic boundaries equals the discharge from the well. Under equilibrium conditions, all pumped water is captured from water that would have discharged to the boundaries in the absence of pumping, or from the boundaries themselves, and none comes from storage.

The first example simulation is of a well pumping from model layer 2 in the general vicinity of Lorella in the upper Lost River subbasin (fig. 48). This location is close to many types of hydrologic boundaries including streams, agricultural drain networks, and evapotranspiration (ET) surfaces. Because of the close proximity of the well to the stream, the pumping affects the stream almost immediately (fig. 49A). The rate of stream impact increases during the pumping season to slightly more than 2 ft³/s, and then diminishes after pumping stops. The maximum rate of stream impact increases slightly during the first few years, but stabilizes after 4 or 5 years of pumping, with the peak impact to streams reaching about 3 ft³/s at the end of the pumping season. Boundaries other than streams are also affected by pumping in this scenario (fig. 49B). Because the well is close to an area of shallow groundwater, discharge to agricultural drains and evapotranspiration by phreatophytes also are reduced by the slight lowering of the water table. The peak impacts to drains and ET are about 0.8 and 0.7 ft³/s respectively at the end of the pumping season (fig. 49B). At equilibrium, the impacts to the various boundaries average 2.5 ft³/s on an annual basis, with the average discharge to streams reduced by 1.7 ft³/s, discharge to drains reduced by 0.6 ft³/s, and ET reduced by 0.2 ft³/s. Impacts to lakes and general head boundaries are less than 0.01 ft³/s and too small to show on a graph. It is important to note that the peak changes in flow to discharge boundaries are a fraction of the 10 ft³/s pumping rate. This is because storage and transmissivity characteristics of the aquifer system buffer the peak pumping and spread the effects out over the entire year.

Because the well in the first example simulation is close to a variety of hydrologic boundaries, the cone of depression does not need to expand far to capture sufficient flow to supply the pumpage (fig. 48). Hence, the drawdown effects are confined to the area close to the well. After 50 years of seasonal pumping, residual drawdown of 2 ft extends only about a mile from the well, and no residual drawdown is indicated farther than 5 to 10 mi from the well. Impacts to streams are concentrated near the pumping well and the reduction in groundwater discharge to streams is as much as 0.4 ft³/s in some model cells (fig. 48).

To demonstrate the effects of pumping intermittently, the first simulation was rerun using an intermittent schedule that repeats 3 years of pumping followed by 3 years of no pumping (fig. 50). It can be seen that the effects of pumping on hydrologic boundaries dissipates almost entirely during the period of no pumping.

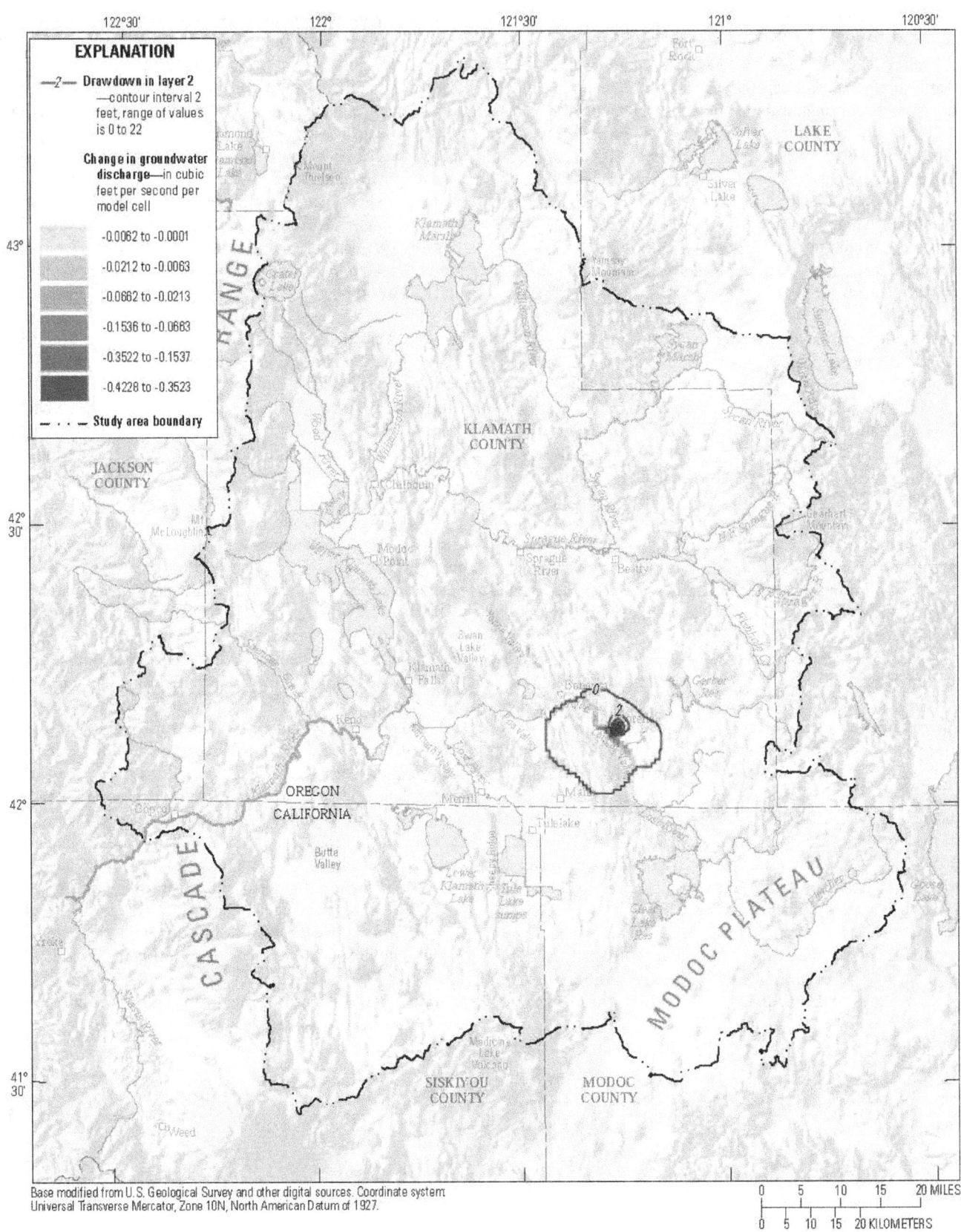

Figure 48.   Drawdown and reductions in groundwater discharge to streams after 50 years of pumping a well in model layer 2 in the upper Lost River subbasin, Oregon, at 10 cubic feet per second for 92 days per year (July–September) each year.

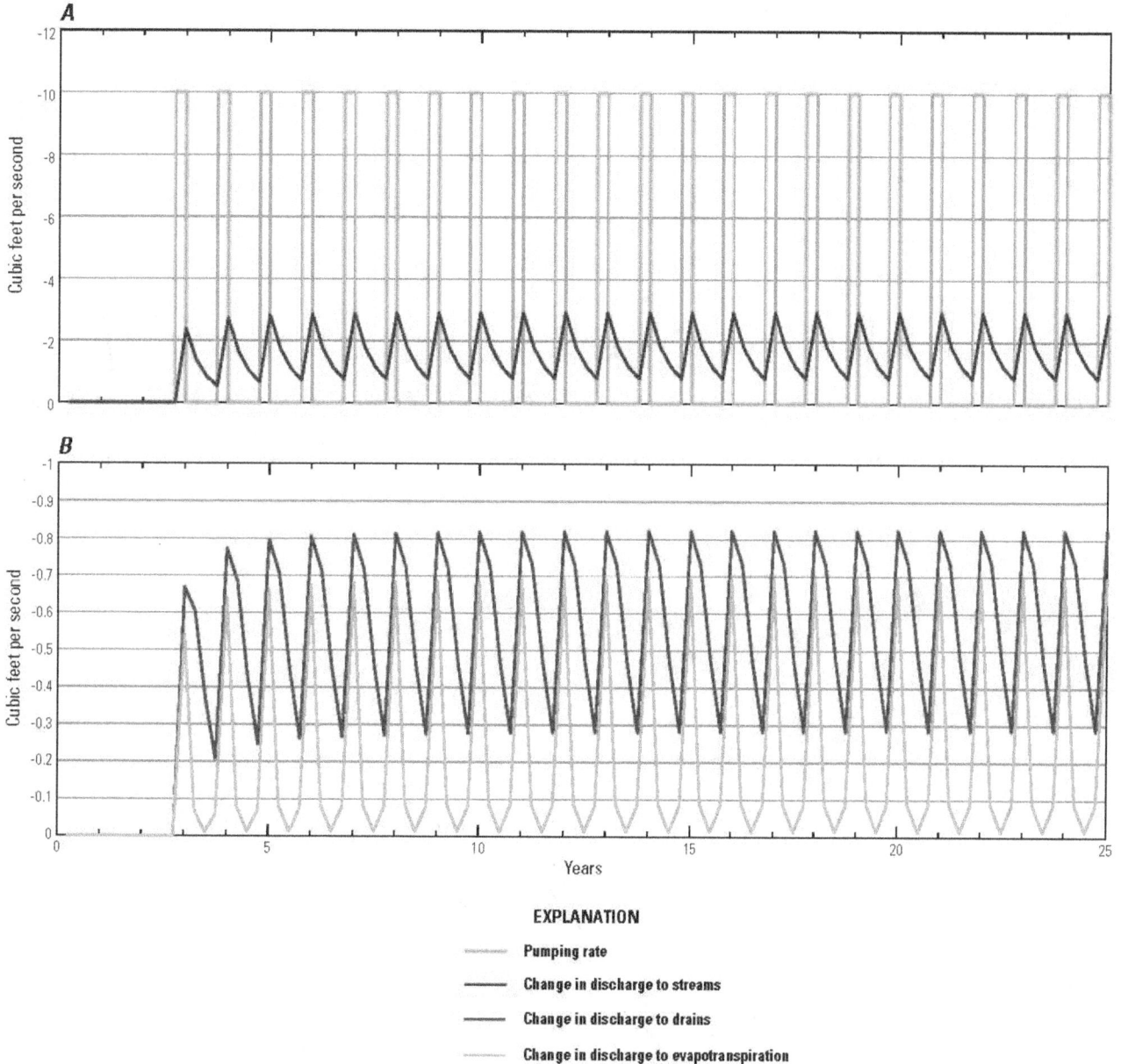

Figure 49. Simulated impacts to hydrologic boundaries due to pumping a well in model layer 2 in the upper Lost River subbasin, Oregon, at 10 cubic feet per second for 92 days per year (July-September) each year. (*A*) Changes in groundwater-discharge rate to streams and (*B*) changes in groundwater-discharge rates to drains and evapotranspiration. Only the first 25 years of the 50-year simulation are shown for clarity.

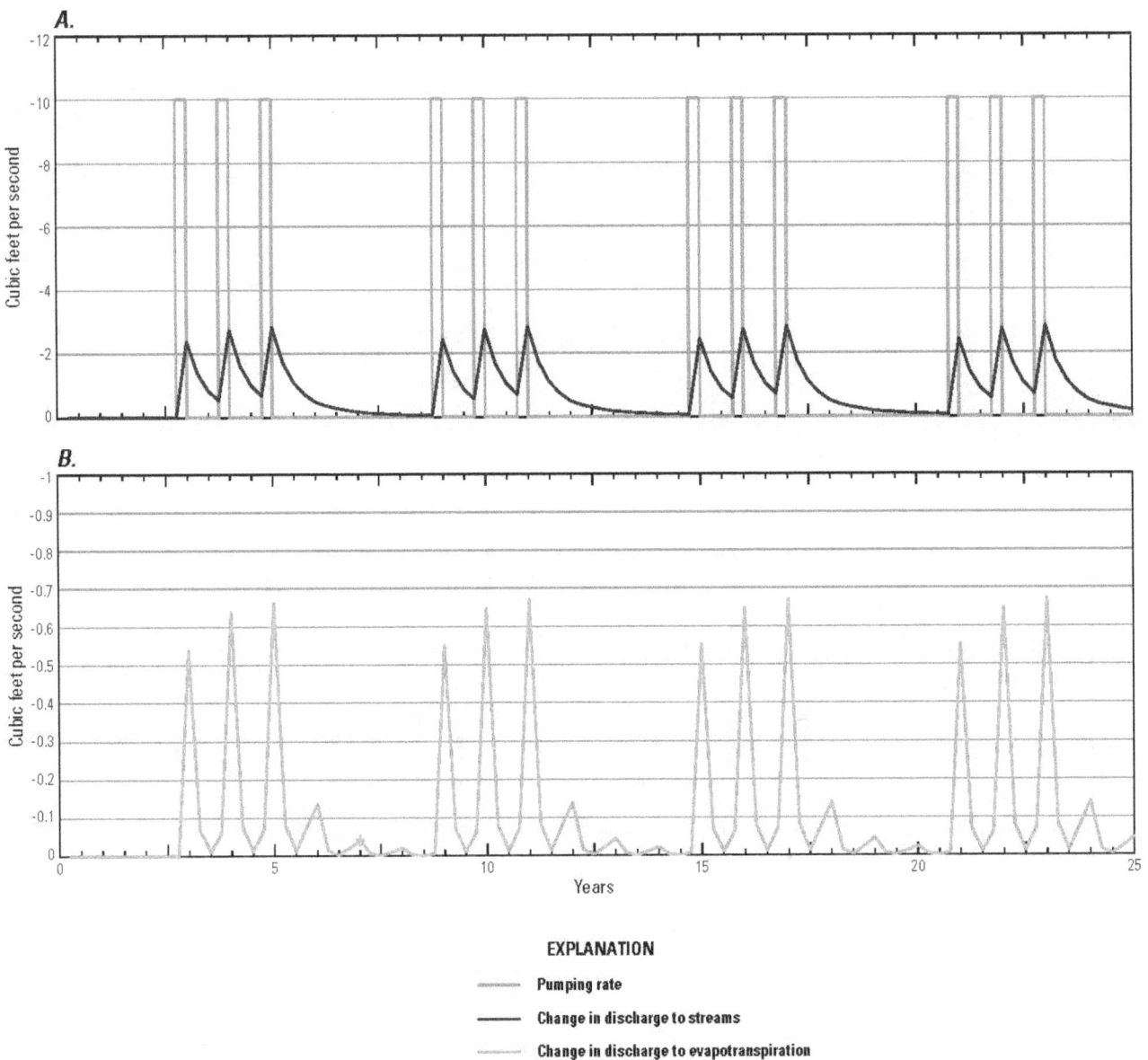

A.

B.

EXPLANATION

- - - - - Pumping rate

———— Change in discharge to streams

- - - - - Change in discharge to evapotranspiration

Figure 50. Simulated impacts to hydrologic boundaries due to pumping a well in model layer 2 in the upper Lost River subbasin at 10 cubic feet per second for 92 days per year (July–September) under a repeating schedule of 3 years with pumping followed by 3 years without pumping. (*A*) Changes in groundwater-discharge rate to streams and (*B*) changes in groundwater-discharge rates to drains and evapotranspiration. Only the first 25 years of the 50-year simulation are shown for clarity.

The second simulation involves a well pumping from model layer 3 about 5 mi south of Beatty, Oregon (fig. 51). This location is farther from groundwater-discharge boundaries in the model (such as streams, extensive agricultural drain networks, and major springs) than in the previous example. Because of this, there is very little effect on hydrologic boundaries at the onset of pumping; nearly all the water pumped by the well is from groundwater storage

for the first few years (fig. 52). The cone of depression expands sufficiently to begin intercepting water discharging to streams after about a year of pumping, but the rates are relatively small for the first few years. Because the well is relatively far from boundaries, the seasonal variations in pumping are almost entirely damped, and the impact to the streams is relatively constant throughout the year (fig. 52).

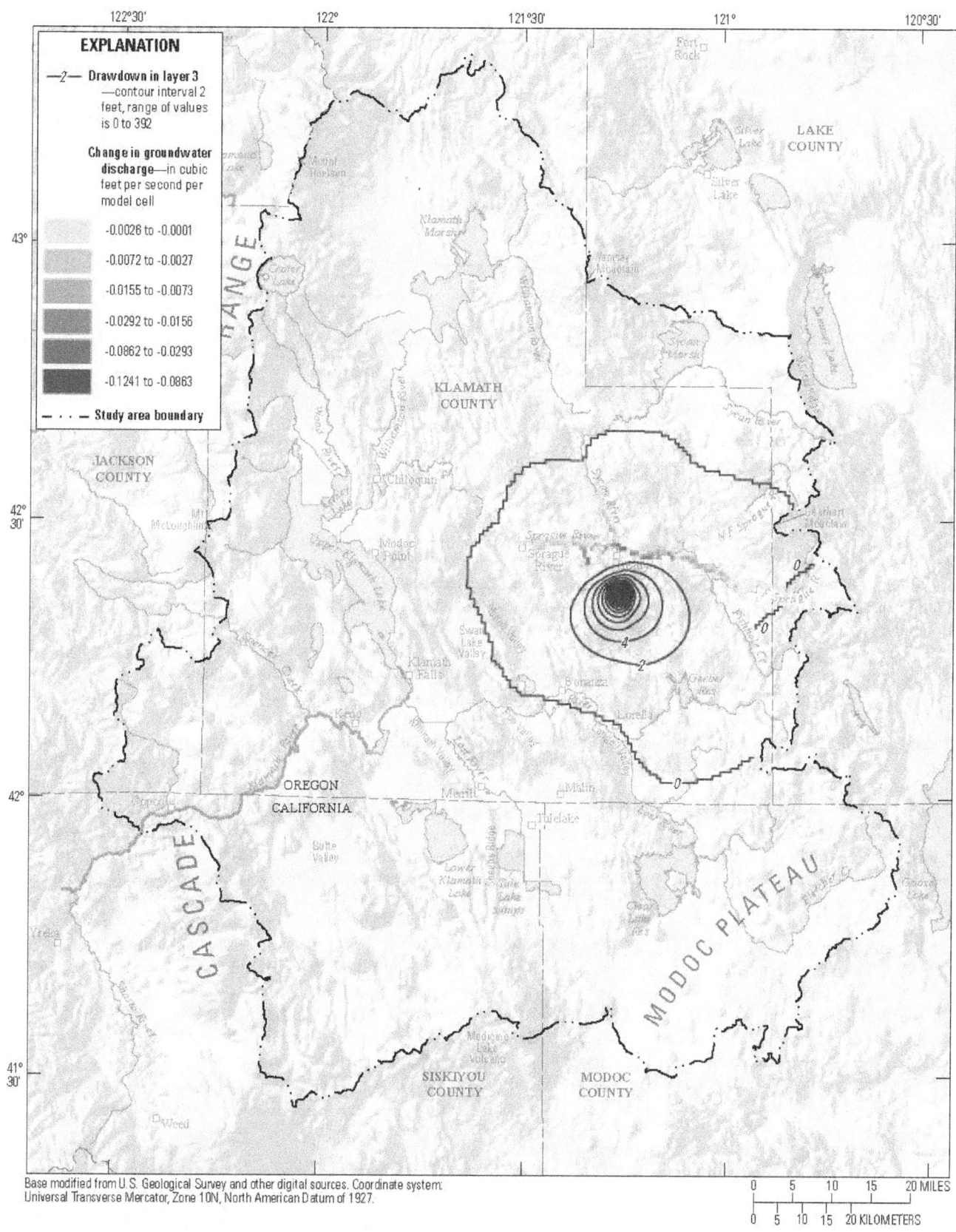

Figure 51. Drawdown and reductions in groundwater discharge to streams after 50 years of pumping a well in model layer 3 approximately 5 miles south of Beatty, Oregon, at 10 cubic feet per second for 92 days per year (July–September) each year.

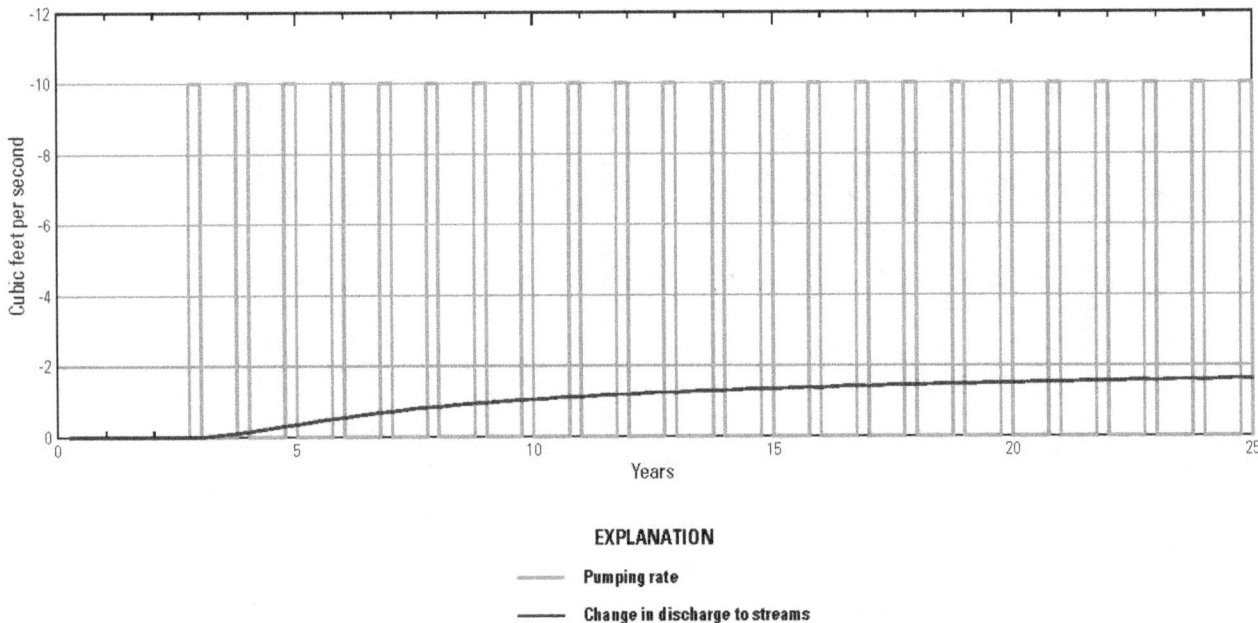

EXPLANATION

——————   **Pumping rate**

——————   **Change in discharge to streams**

Figure 52.   Simulated impacts to hydrologic boundaries due to pumping a well in model layer 3 approximately 5 miles south of Beatty, Oregon, at 10 cubic feet per second for 92 days per year (July–September) each year. Only the first 25 years of the 50-year simulation are shown for clarity.

Additionally, because of the large distance from boundaries, the cone of depression is slow to stabilize and does not reach equilibrium during the simulation. At the end of the 50-year simulation period, about 1.8 ft³/s of the average yearly pumping is being captured from diminished discharge to streams, and about 0.6 ft³/s is still coming from groundwater storage; impacts to all other boundaries total about 0.1 ft³/s.

Because the well in this second simulation is distant from hydrologic boundaries, the cone of depression spreads over a considerable area before capturing sufficient flow to supply the pumpage (fig. 51). At the end of the simulation period, residual drawdown of 2 ft extends as much as 10 mi from the wells, and drawdowns between 0 ft and 2 ft are simulated as far as 20 mi. Because of the large area covered by the cone of depression, stream impacts are spread over a broad area, but are generally small in any given stream reach (fig. 51). The largest reduction of groundwater discharge to streams in any model cell is about 0.12 ft³/s.

The third simulation involves a well pumping from model layer 3 near the center of the Tule Lake subbasin (fig. 53). This well is close to an extensive network of agricultural drains, an area of shallow groundwater (and hence, phreatophytic plants), and the Tule Lake sumps, which are simulated in a manner similar to lakes. Although a stream is close to the well (the

lower Lost River), it is not a major location of groundwater discharge. In this simulation, the cone of depression appears to stabilize within several years, and impacts to hydrologic boundaries reach a steady state (fig. 54). Most of the pumped water is captured from diminished discharge to agricultural drains, with smaller amounts captured from reduced ET and net discharge to lakes (primarily the Tule Lake sumps in this case). The interaction between time varying ET rates and drains cause the peak drain impacts (which are just below 2 ft³/s) to occur after the end of the pumping season. Peak reductions in ET of about 0.8 ft³/s occur at the end of the pumping season. The reduction in net groundwater discharge to lakes is about 0.36 ft³/s, with almost no seasonal variation. The largest reductions in groundwater discharge to streams are less than 0.005 ft³/s per model cell. Because the combined impacts to streams and general head boundaries total to less than 0.1 ft³/s they are not shown on figure 54.

The hydrologic characteristics of the deep aquifer (model layer 3) in the area, and the distribution of hydrologic boundaries result in a fairly flat cone of depression that spreads out about 15 mi from the pumping well (fig. 53). Drawdown is less than 2 ft at the end of the 50-year simulation period except within about 1.5 mi of the well.

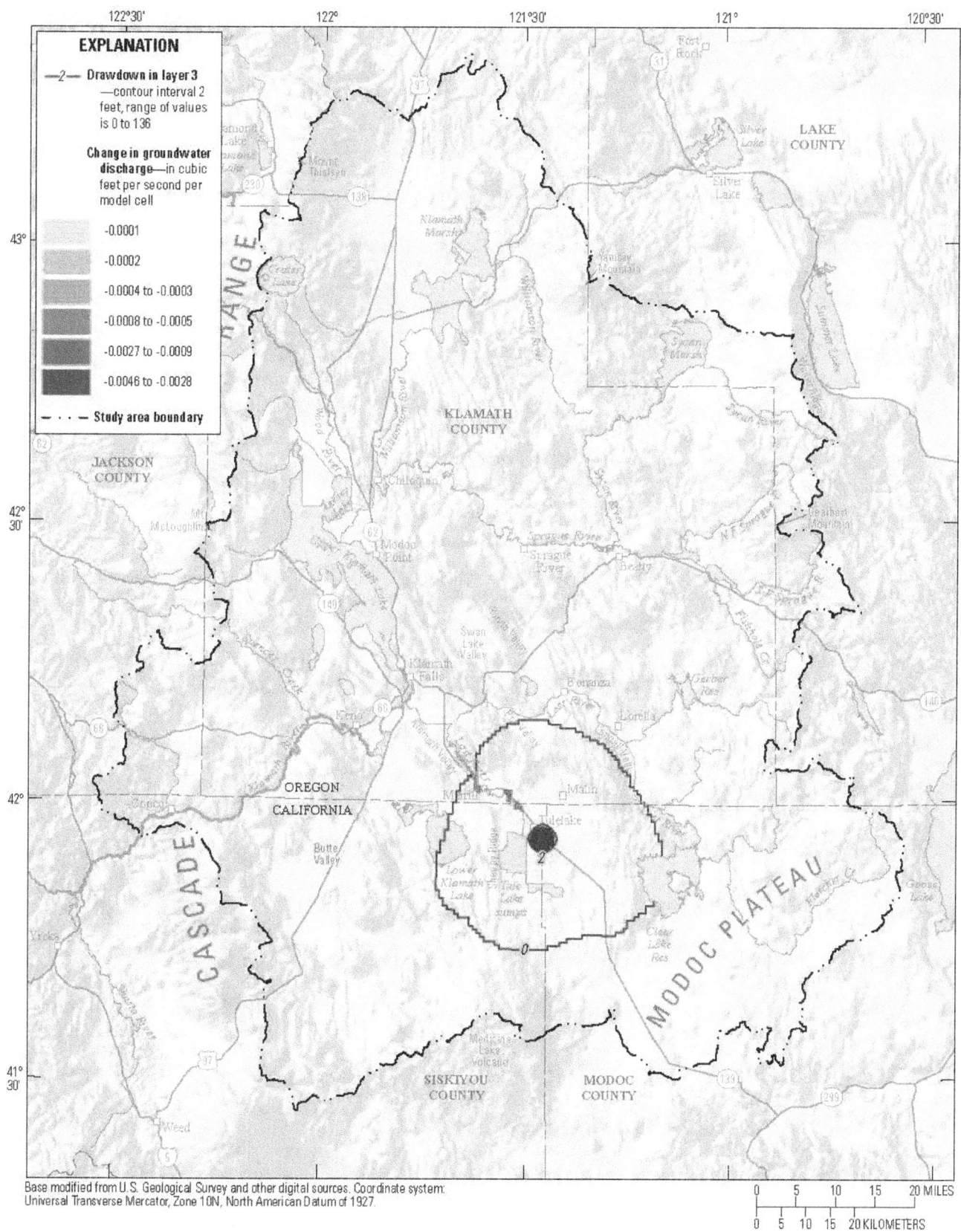

Figure 53. Drawdown and reductions in groundwater discharge to streams after 50 years of pumping a well in model layer 3 near the middle of the Tule Lake subbasin, Oregon and California, at 10 cubic feet per second for 92 days per year (July–September) each year.

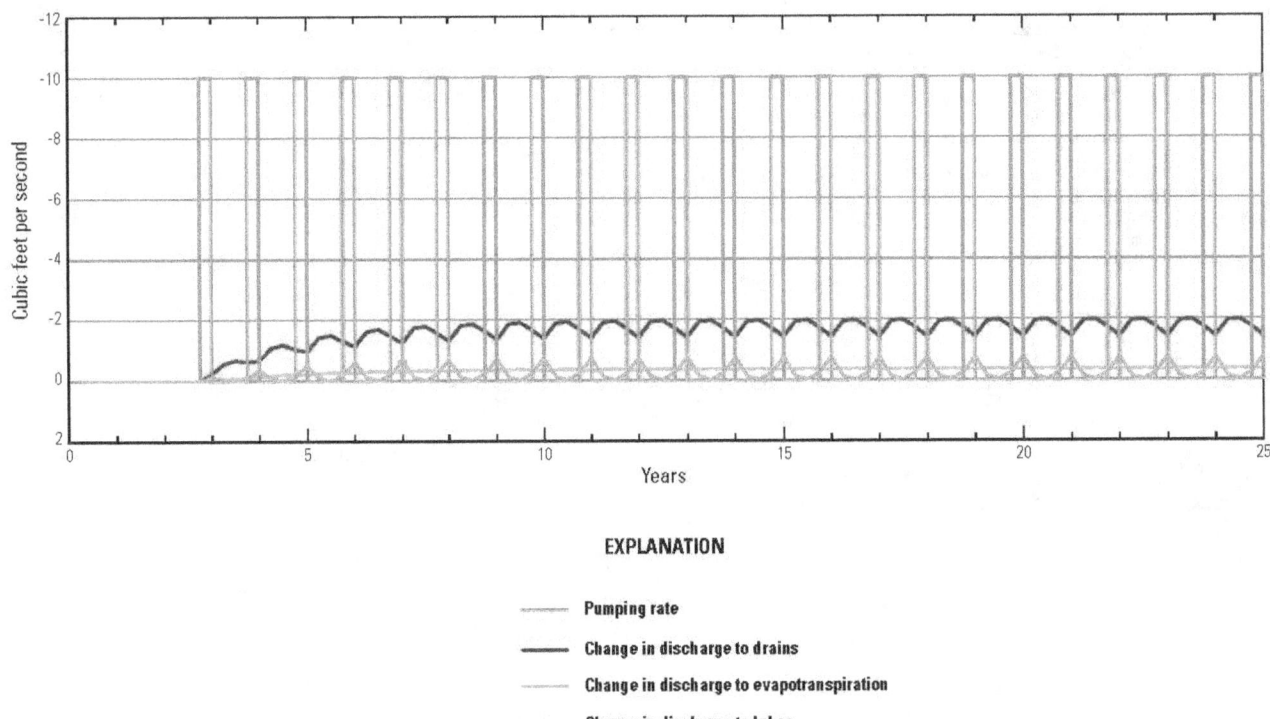

EXPLANATION

—————— Pumping rate

—————— Change in discharge to drains

—————— Change in discharge to evapotranspiration

—————— Change in discharge to lakes

Figure 54.   Simulated impacts to hydrologic boundaries due to pumping a well in model layer 3 near the center of the Tule Lake subbasin, Oregon and California, at 10 cubic feet per second for 92 days per year (July–September) each year.

Groundwater Management Model

Water-Management Issues

Water resources in the upper Klamath Basin are managed to achieve a variety of complex and interconnected purposes. Over the past decade, balancing the benefits of water for agriculture and for ecological needs has proven difficult. A series of Endangered Species Act biological opinions have required the Bureau of Reclamation's Klamath Project to limit diversions from Upper Klamath Lake and the Klamath River (the principal sources of Project irrigation water) in order to protect habitat for endangered and threatened fish. Since 2001, Reclamation has been required to maintain elevations in Upper Klamath Lake to protect habitat for the endangered Lost River sucker (*Deltistes luxatus*) and shortnose sucker (*Chasmistes brevirostris*) (U.S. Fish and Wildlife Service, 2008), while simultaneously providing specified flows in the Klamath River to protect habitat for the coho salmon federally listed as threatened (National Marine Fisheries Service, 2010). This shift in water-management priorities resulted in substantial reductions in the amount of surface water diverted to the Project in 2001 and 2010, and has increased the likelihood that the Project will face water shortages in the future.

In response to changing water-management priorities, Klamath Basin stakeholders have developed the proposed Klamath Basin Restoration Agreement (KBRA), which aims to restore historic fish habitat and populations in the upper Klamath Basin and establish reliable water supplies for agriculture (Klamath Basin Restoration Agreement, 2010). The KBRA includes a number of actions that substantially change water-resources management in the basin. Among those is the Water Resources Program (KBRA, Part IV), which establishes a permanent limitation on the amount of water that will be diverted from Upper Klamath Lake and the Klamath River. The proposed limitations on diversion, which are based on the forecast for net inflow to Upper Klamath Lake during the period April 1–September 30, vary from 330,000 acre-ft in low-inflow years to a maximum of 385,000 acre-ft in high-inflow years. An additional 48,000 acre-ft (low-inflow year) to 60,000 acre-ft (high-inflow year) will be diverted to the Lower Klamath Lake and Tule Lake National Wildlife Refuges, resulting in total annual maximum diversions of 378,000 to 445,000 acre-ft (KBRA, Section 15 and appendix E-1). The Water Resources Program also includes actions to improve streamflows and maintain the elevation of Upper Klamath Lake. The Agreement's Water Use Retirement Program will rely on voluntary retirement of water rights or water uses to secure 30,000 acre-ft of water to increase inflow into Upper Klamath Lake (KBRA, Section 16), and the On-Project Plan includes criteria to ensure groundwater development does not have significant impacts on essential environmental flows (KBRA, Section 15).

In contrast to the water management plan defined by the KBRA, historical water diversions to meet Project and refuge needs have actually been largest during dry years when inflows to Upper Klamath Lake tended to be smaller than average. This variability in diversions is primarily due to climate and reflects the increased demand for water in dry years. Prior to 2001, annual diversions from Upper Klamath Lake and the Klamath River (for both Klamath Project and refuge needs) ranged from about 320,000 to about 490,000 acre-ft (McFarland and others, 2005). A comparison of historical water diversion amounts to the maximum diversion amounts proposed in the KBRA indicates a reduction in the amount of water that can be diverted to the Project in dry years of as much as approximately 100,000 acre-ft.

Shifting water-management priorities are expected to create a sustained demand for groundwater to supplement surface-water supplies. Since 2001, groundwater use in the upper Klamath Basin has increased substantially, largely due to programs funded by Reclamation to augment the Klamath Irrigation Project's surface-water supplies. These programs used a variety of methods in which farmers were compensated to use pumped groundwater as a substitute for surface water. In 2004, groundwater development associated with these programs increased groundwater use in the basin by about 50 percent over the estimated pumping in 2000, and more than doubled groundwater use in the area of the Bureau of Reclamation Klamath Project (Gannett and others, 2007). This sharp increase in pumping resulted in groundwater-level declines of 10 to 15 ft over much of the Project area. If groundwater use continues at the current level, the groundwater system will eventually achieve a new state of dynamic equilibrium. However, the spatial and temporal impact of this increased pumping on groundwater and surface water is unknown. The goal of the optimization model presented below is to assess the effect of sustained groundwater pumping on the complex and interconnected groundwater and surface-water system of the upper Klamath Basin in order to identify groundwater development strategies that result in acceptable impacts to groundwater and surface-water resources.

Approach for Evaluating Future Groundwater Development

Groundwater development alternatives for the upper Klamath Basin were evaluated with the aid of coupled groundwater simulation and optimization models. A series of simulation-optimization models were developed to identify the key features and limitations of future groundwater development in the basin. The simulation-optimization models were designed to include the important hydrologic characteristics of groundwater development, such as the link between groundwater withdrawal and groundwater discharge to streams and lakes, the effect of increased pumping on existing groundwater users, and the impact of groundwater withdrawals on the Project's drain system.

An important issue in testing future groundwater-development scenarios is the impact of increased pumping on groundwater discharge that supports wildlife habitat, out-of-stream uses, and scenic flows. A substantial proportion of streamflow in the upper Klamath Basin consists of groundwater discharge, particularly in the late-summer months when streamflow is needed to meet critical environmental needs for endangered and threatened fishes. The KBRA provides measures for defining and limiting the effects of groundwater pumping on environmental flows. The KBRA defines the "adverse impact" of pumping to mean a 6 percent or greater reduction in groundwater discharge to springs associated with Upper Klamath Lake, the Wood River and its tributaries, Spring Creek, the lower Williamson River, and the Klamath River and its tributaries. The KBRA further defines the measure of "adverse impact" to be only those effects of groundwater withdrawal that are caused by groundwater use within the Project. That is, the impact of groundwater pumping for the purpose of augmenting the Project's surface-water supplies will be measured independently of the impact of all other stresses active in the basin. The simulation-optimization model presented below evaluates all groundwater-development scenarios to ensure that the impacts of groundwater withdrawal on environmental flows are within the limits defined in the KBRA.

A second issue to consider is the impact of groundwater development on existing groundwater users. OWRD determines whether wells have the potential to impair other groundwater users by causing substantial interference, a term that comprises many factors but generally refers to declines in groundwater levels that impair the ability of those with senior water rights to pump groundwater (Oregon Administrative Rules, 2011). Historically, substantial interference with other groundwater users has rarely been a problem in the upper Klamath Basin. However, due to the unprecedented increase in groundwater use since 2000, OWRD has been conditioning most new water right permits with water-level decline constraints. The conditions address short term (seasonal), year-to-year, and long term (multi-year) water-level declines. The simulation-optimization model has the capacity to test alternative groundwater-development strategies to ensure that drawdowns at a range of time scales are controlled and do not exceed limits consistent with OWRD standards.

Finally, the analysis of groundwater-development scenarios must consider the impact of groundwater development on the operation of the Project. Subregions of the Project are hydrologically connected through the natural river network and the system of canals and drains. There is extensive re-circulation of drainage water throughout the Project, and the majority of on-farm irrigation inefficiencies are recycled through the use of drain water for irrigation (Burt and Freeman, 2003). Consequently, any groundwater discharge diverted from the drain system will reduce the amount of water available for irrigation. The Project's infrastructure does not provide detailed measurement of the spatial and temporal variations of drain discharge or a detailed assessment of the spatial and temporal demand for drain water. Although there are no guidelines for defining excessive depletions in groundwater discharge to the Project's drain system, the optimization model provides a flexible framework for evaluating a varied set of depletion limits.

Simulation-Optimization Analysis

The groundwater flow model previously described in this report was coupled with techniques of constrained optimization to evaluate alternative groundwater-development plans for the upper Klamath Basin. The example simulation-optimization model described here consists of a mathematical formulation of groundwater-development goals (objective function) and a set of example constraints that limit those goals. The constraints are included to incorporate physical limitations of the groundwater system, along with other performance criteria, that must be honored. It should be noted that the example in this report does not incorporate climate (drought) cycles and off-project supplemental groundwater pumping, both of which will affect results. In addition, the example constraints may not represent those chosen by resource managers and water users in actual practice. Consequently, the example presented in this section is intended to demonstrate the concepts, functioning, and utility of the simulation-optimization model, and the numerical results are not intended to represent actual pumping targets. The simulation-optimization model:

- Evaluates groundwater/surface-water interactions for streams and lakes throughout the basin and ensures that the impacts of Klamath Project pumping on surface-water resources are within prescribed limits.

- Assesses the impact of Project pumping on groundwater levels and ensures that short- and long-term drawdowns do not exceed limits designed to protect other groundwater users.

- Identifies groundwater-development strategies for the Project that limit the impact of groundwater pumping on the Klamath Project drain system.

- Identifies groundwater-withdrawal strategies that support seasonal water demands for irrigation within the Project.

Mathematical Formulation of Objective and Constraints

The groundwater-optimization model applies the technique of constrained optimization, which uses mathematically formalized checks and balances to integrate complex water-management issues and quantitatively compare alternative groundwater-development plans. The use of optimization techniques allows consideration of an essentially limitless number of groundwater-development options. Formulation of the optimization model involves defining the model objective, decision variables, and a set of constraints. The optimization technique used in this study is sequential linear programming. The objective equation is linear; all constraint equations are linear or are linearized to allow solution using the sequential linear programming technique. Decision variables represent quarterly pumping at managed wells and are determined by the optimization model. The constraint set places limitations on drawdown and reduction in groundwater discharge to surface water and drains. The constraint set also places limitations on groundwater pumping to define seasonal demands on withdrawal rates and to define the allowable range of withdrawal for each well.

Groundwater has been pumped at public-supply and irrigation wells throughout the upper Klamath Basin. The groundwater model described in this report includes about 1,000 wells that have been used during the period 1970–2004. The managed wells used in the optimization model are shown in figure 55A and represent the wells used in Reclamation's groundwater acquisition program and pilot water bank. The optimization model calculates the pumping rates for the 112 managed wells; all other wells in the model are set to water-year 2000 pumping rates and are considered to be background stresses that contribute to the total stress acting on the groundwater system, but do not change from one simulation to the next.

The objective function of the optimization model was formulated to maximize the total pumping from the managed wells and is given by

$$maximize \sum_{i=1}^{NW} \sum_{j'=3}^{4} Qw_{i,j'}, \qquad (7)$$

where
$Qw_{i,j'}$ is the pumping rate at well i in water-year quarter j', and
NW is the number of managed wells and equals 112.

The objective function sums the quarterly pumping rate at each managed well over water-year quarters 3 and 4, which define the April–October irrigation season. The optimization model has 224 decision variables, which are the 3rd- and 4th-quarter pumping rates for each of the 112 managed wells.

The analysis of groundwater-withdrawal strategies requires measures, known as constraints, to define the success or failure of alternative plans. The pumping rates identified by the simulation-optimization models were limited by a set of specified constraints on short- and long-term drawdowns; reduction in groundwater discharge to streams, lakes, and drains; seasonal demand for groundwater; and withdrawal rates for each well. The mathematical formulations of these constraints are defined in the following sections. The constraints used in a specific formulation of the optimization model, along with the constraint limits, varied among the different model applications.

Groundwater withdrawal can have both short- and long-term impacts on groundwater levels. These range from the steep/large drawdown that generally occurs close to a withdrawal well and has a rapid onset and recovery, to the long-term drawdown that typically develops over a broad region after a period of sustained pumping. The optimization model included constraints on drawdown at 438 drawdown control sites shown in figure 55B. For each control location, three types of drawdown constraint—seasonal, year-to-year, and decadal—were defined. The numerical values used in this test case are broadly consistent with OWRD standards. The 2,500 ft model grid spacing limits the ability of the model to simulate pumping effects at finer spatial scales. As such, the model may underestimate large drawdowns very close to actively pumping wells. For this reason, seasonal-drawdown constraints were set to limits slightly smaller than presently used for regulatory purposes. First, seasonal-drawdown constraints were defined to limit the lowering of the groundwater levels during the irrigation season (fig. 56).

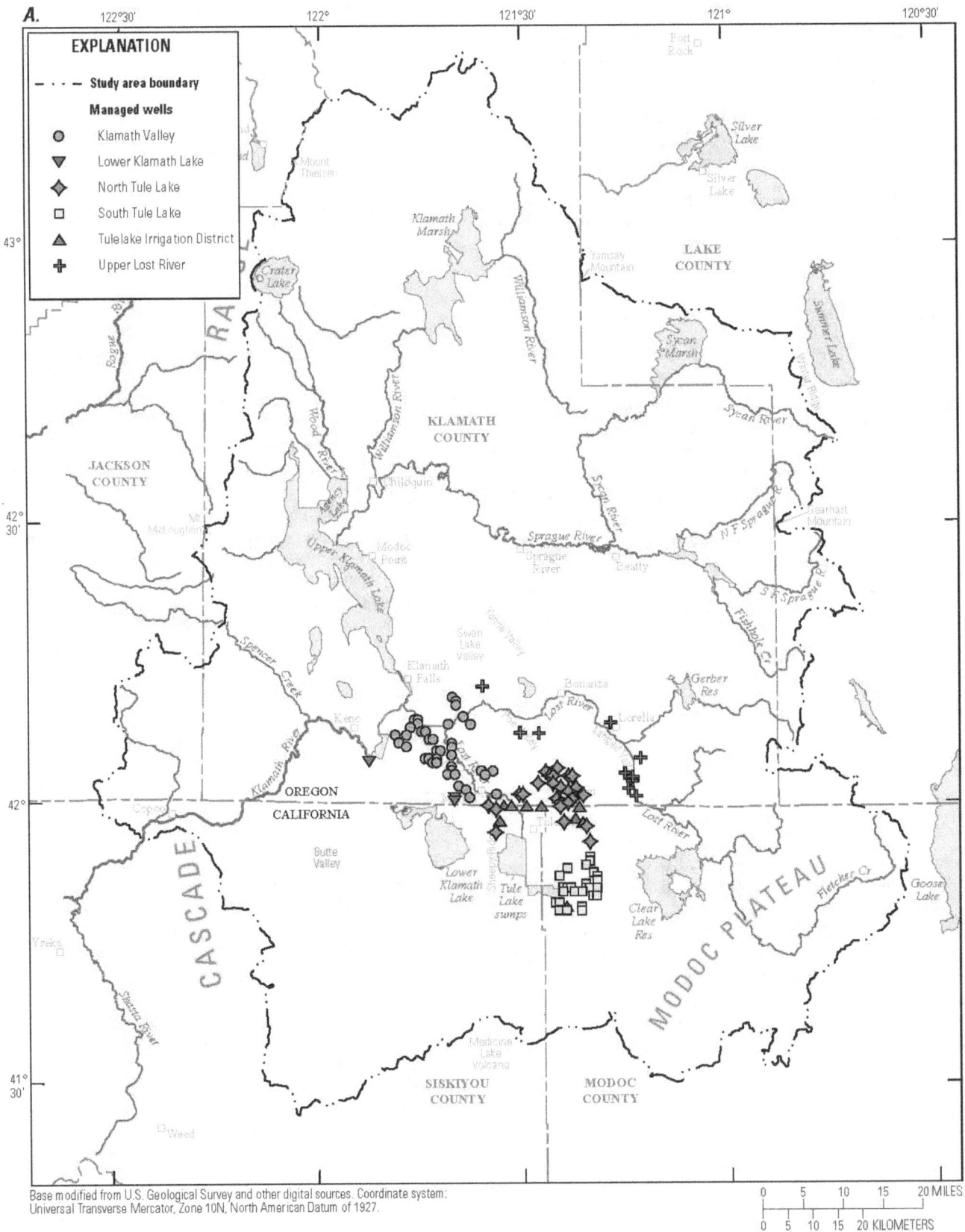

Figure 55.   Upper Klamath Basin, Oregon and California, with locations of (*A*) optimized wells and (*B*) drawdown and discharge constraints.

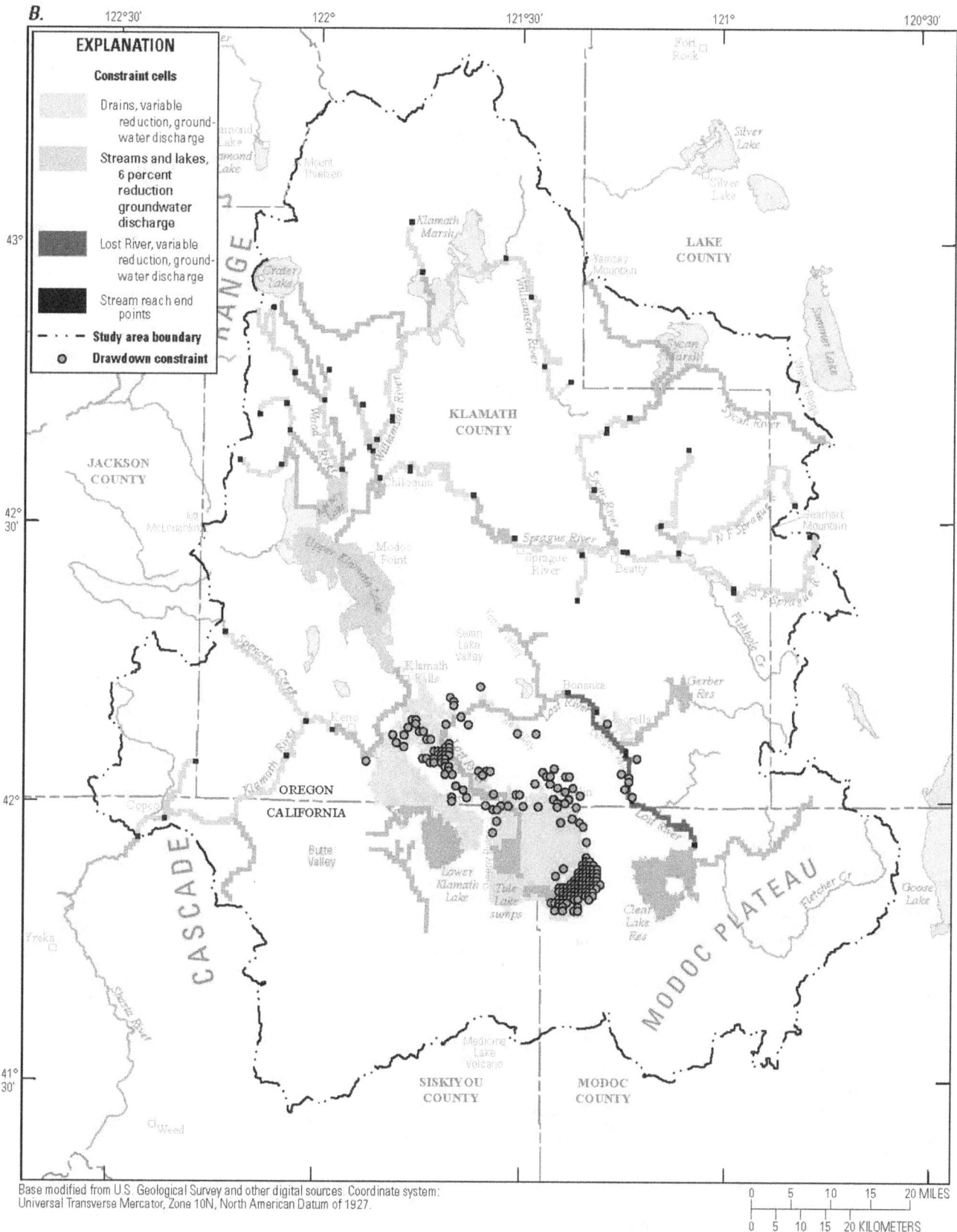

B.

EXPLANATION

Constraint cells

Drains, variable reduction, groundwater discharge

Streams and lakes, 6 percent reduction groundwater discharge

Lost River, variable reduction, groundwater discharge

Stream reach end points

—··— **Study area boundary**

◉ **Drawdown constraint**

Base modified from U.S. Geological Survey and other digital sources. Coordinate system: Universal Transverse Mercator, Zone 10N, North American Datum of 1927.

0 5 10 15 20 MILES

0 5 10 15 20 KILOMETERS

Figure 55.—Continued

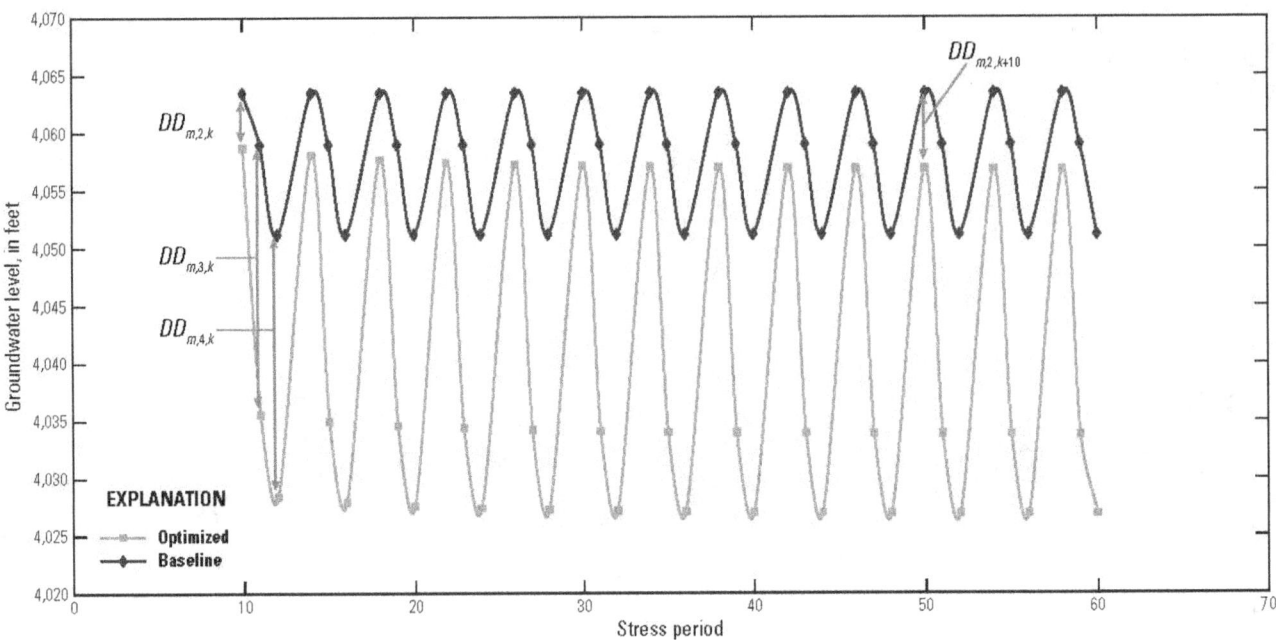

Figure 56. Example of baseline and optimized hydraulic heads showing values used in the definition of drawdown constraints (equations 8 and 9).

$$DD_{m,3,k} - DD_{m,2,k} \leq DD_{seas,max}, \qquad (8)$$

$$DD_{m,4,k} - DD_{m,2,k} \leq DD_{seas,max}, \qquad (9)$$

where

$DD_{m,2,k}$ is the drawdown at constraint location m at the end of quarter 2 (start of irrigation season) in water year k,

$DD_{m,3,k}$ is the drawdown at constraint location m at the end of quarter 3 (midpoint of irrigation season) in water year k,

$DD_{m,4,k}$ is the drawdown at constraint location m at the end of quarter 4 (end of irrigation season) in water year k, and

$DD_{seas,max}$ is the upper bound on seasonal drawdown.

As illustrated in figure 56, seasonal drawdowns defined in equations 8 and 9 ($DD_{m,2,k}$, $DD_{m,3,k}$, and $DD_{m,4,k}$) are determined with respect to the water level calculated at the end of each quarter for the background simulation; that is, the simulation without managed pumping. The drawdowns defined in equations 8 and 9 are not the same as the change in groundwater levels calculated for the optimal pumping strategies from the second to the third (or second to fourth) water-year quarters. The seasonal drawdown constraints seek to limit only the drawdown that occurs due to the supplementary groundwater pumping evaluated in the optimization model, and therefore must take into account the variation in groundwater levels that occurs in response to the background stresses, as defined in equations 8 and 9

and presented in figure 56. The seasonal-drawdown limit was initially set to 20 ft. The impact of this limit was evaluated through sensitivity analysis.

A second, year-to-year drawdown constraint was defined to limit the "residual" drawdown remaining at the beginning of the irrigation season due to the previous year's withdrawal.

$$DD_{m,2,k+1} - DD_{m,2,k} \leq DD_{year,max}, \qquad (10)$$

where

$DD_{m,2,k+1}$ is the drawdown at constraint m at the end of quarter 2 of water year $k+1$, and

$DD_{year,max}$ is the upper bound on year-to-year drawdown.

The year-to-year drawdown limit was initially set to 4 ft. A sensitivity analysis tested the impact of this constraint on model results.

A third, decadal drawdown constraint was included to limit the long-term drawdown resulting from sustained pumping over many years. Long-term drawdown was defined as the drawdown that occurs over a 10-year period.

$$DD_{m,2,k+10} - DD_{m,2,k} \leq DD_{10yr,max}, \qquad (11)$$

where

$DD_{m,2,k+10}$ is the drawdown at constraint location m at the end of quarter 2 of water year $k+10$, and

$DD_{10yr,max}$ is the upper bound on drawdown over a 10-year period.

The long-term drawdown limit was set to 25 ft. Preliminary analyses indicated that the groundwater management model solution was not sensitive to this constraint for long-term drawdown limits greater than or equal to 10 feet. Therefore, this constraint limit remains unchanged in all model runs.

Groundwater withdrawal is accompanied by declines in water levels that interact with surface water at a variety of temporal and spatial scales, potentially causing reductions in groundwater discharge to streams, lakes, and drains. Groundwater discharge to surface water is an important component of environmental flows that support wildlife habitat in the upper Klamath Basin, and groundwater discharge to drains supplies a component of the Project's irrigation-water needs. Constraints on the depletion of groundwater discharge to streams and lakes required the depletion to be less than or equal to a specified maximum for 32 stream reaches throughout the basin and Upper Klamath Lake (fig. 55B).

$$QSR_{m,j,k} \leq QSR_{m,j,k,max}, \qquad (12)$$

$$QLR_{ukl,j,k} \leq QLR_{ukl,j,k,max}, \qquad (13)$$

where

$QSR_{m,j,k}$ is the reduction in groundwater discharge to stream reach m in quarter j of water year k,

$QSR_{m,j,k,max}$ is the upper bound on reduction in groundwater discharge to reach m in quarter j of water year k,

$QLR_{ukl,j,k}$ is the reduction in groundwater discharge to Upper Klamath Lake in quarter j of water year, k, and

$QLR_{ukl,j,k,max}$ is the upper bound on reduction in groundwater discharge to Upper Klamath Lake in quarter j of water year k.

Constraints on the reduction in groundwater discharge to the Klamath River and Upper Klamath Lake and their tributaries were set to 6 percent of the baseline groundwater discharge, as specified in the KBRA, and were fixed at this value in all model runs. The base-case optimization analysis also includes a 6-percent depletion limit for the Lost River. Alternative formulations of the optimization model tested the impact of the Lost River constraint limit on model results.

Similar constraints were defined to limit the reduction in groundwater discharge to drains (fig. 55B).

$$QDR_{j,k} \leq QDR_{j,k,max}, \qquad (14)$$

where

$QDR_{j,k}$ is the cumulative reduction in groundwater discharge to the entire drain system in quarter j of water year k, and

$QDR_{j,k,max}$ is the upper bound on reduction in groundwater discharge to drains in quarter j of water year k.

The constraints on the reduction in groundwater discharge to drains were initially set to 20 percent of the baseline discharge. The impact of these constraints was tested using sensitivity analysis.

A demand constraint was included to evaluate the impact of seasonal demand for groundwater. The demand constraint imposes a minimum value on the sum of 4th quarter withdrawal rates from all wells.

$$\sum_{i=1}^{NW} Qw_{i,4} \geq Qw_{4,min}, \qquad (15)$$

where

$Qw_{i,4}$ is the withdrawal rate at well i in water-year quarter 4, and

$Qw_{4,min}$ is the lower bound on total withdrawal across all wells in water-year quarter 4.

The base-case optimization model formulation does not include a demand constraint; a sensitivity analysis was performed to test the impact of this constraint.

Finally, constraints on the minimum and maximum withdrawal rates for each well were defined as

$$0 \leq Qw_i \leq Qw_{i,max}, \qquad (16)$$

where

$Qw_{i,max}$ is the upper bound on withdrawal rate for well i.

Upper bounds on withdrawal rates (maximum pumping rates) were estimated from water bank pumping records by dividing reported pumped volumes by the reporting period. Because this method assumes the volume was produced by continuous pumping over the reporting period, the estimated pumping rates are conservative. This is generally not a problem for this test case because estimated maximum pumping capacities were seldom reached in any of the solutions.

In summary, the groundwater management model was defined to maximize withdrawals from managed wells (equation 7), subject to constraints on drawdown (equations 8–11); reductions in groundwater discharge to streams, lakes, and drains (equations 12–14); water demand (equation 15); and withdrawal rates at managed wells (equation 16). Constraints on seasonal, year-to-year, and long-term drawdown were initially set to 20 ft, 4 ft, and 25 ft, respectively; the impacts of these limits on total withdrawal were evaluated through sensitivity analyses. Constraints on the depletion of groundwater discharge to the Klamath River and Upper Klamath Lake and its tributaries were fixed at 6 percent of baseline discharge values, as specified in the KBRA. Reductions in groundwater discharge to the Lost River were initially limited to 6 percent of baseline discharge values; alternative formulations tested the impact of varying the Lost River depletion limits. The depletion of groundwater discharge to drains within the Project was initially limited to 20 percent of baseline discharge, and was subsequently varied as part of a sensitivity analysis. Finally, the 4th quarter pumping demand constraint was set to zero in the base-case formulation, with alternative formulations tested in a sensitivity analysis.

Response-Matrix Technique

The groundwater management model was formulated using the widely applied response-matrix technique. In this technique, unit solutions to the governing groundwater-flow equation (equation 1) are developed and linearly superposed to simulate the effect of groundwater withdrawal on drawdown and groundwater discharge. The responses are calculated only for the values of interest (drawdown and groundwater discharge to streams, lakes, and drains at constraint sites shown in figure 55) and only as a function of withdrawal at the managed wells shown in figure 55. The responses are compiled in a response matrix that is included as a set of constraints in the optimization model. The result is a compact version of the groundwater model that simulates the effect of managed withdrawal on water levels and groundwater discharge at those locations of critical interest to water-resources management. Detailed developments of the response-matrix technique can be found in Gorelick and others (1993) and Ahlfeld and Mulligan (2000). Literature reviews of simulation-optimization research can be found in Gorelick (1983), Yeh (1992), and Wagner (1995). Example applications that utilize optimization methods to address issues of groundwater development and groundwater–surface-water interactions include Barlow and Dickerman (2001), DeSimone and others (2002) and Granato and Barlow (2004).

The first step in implementing the response-matrix technique is calculation of the drawdown or groundwater-discharge responses (at each of the constraint sites shown in fig. 55B) to simulated unit pumping rates at each of the wells in figure 55A. Calculation of the drawdown and discharge responses required $2NW + 1$ simulations of the transient simulation model, where NW is the number of managed wells in the groundwater management model objective function (equation 7). In each of the $2NW$ simulations, the withdrawal rate for well i in water-year quarter j' was increased by the unit pumping rate, $\Delta Qw_{i,j'}$; at the end of the quarterly stress period, the withdrawal rate for well i was returned to zero. The drawdown or reduction in groundwater discharge resulting from the unit withdrawal was determined by subtracting hydraulic heads (or groundwater discharge) simulated with the unit withdrawal rate from those simulated with background conditions in which the unit rate is inactive. The drawdown at constraint site m in quarter j of water year k, caused by a unit withdrawal at well i in quarter j', is defined as $dd_{m,j,k,i,j'}$. Drawdown response coefficients, $rdd_{m,j,k,i,j'}$, are then defined as

$$rdd_{m,j,k,i,j'} = \frac{dd_{m,j,k,i,j'}}{\Delta Qw_{i,j'}}. \tag{17}$$

The characteristic drawdown responses are recorded for water-year quarters $j = 2,3,4$ (the responses needed in the calculation of drawdown constraints) due to irrigation-season pumping ($j' = 3,4$).

Similarly, the groundwater-discharge responses to a unit withdrawal at well i are defined as

$$rqs_{m,j,k,i,j'} = \frac{qsr_{m,j,k,i,j'}}{\Delta Qw_{i,j'}}, \tag{18}$$

$$rql_{ukl,j,k,i,j'} = \frac{qlr_{ukl,j,k,i,j'}}{\Delta Qw_{i,j'}}, \tag{19}$$

$$rqd_{j,k,i,j'} = \frac{qdr_{j,k,i,j'}}{\Delta Qw_{i,j'}}, \tag{20}$$

where

$qsr_{m,j,k,i,j'}$ is the reduction in groundwater discharge to stream reach m in water-year quarter j of year k caused by a unit withdrawal at well i in water-year quarter j',

$qlr_{ukl,j,k,i,j'}$ is the reduction in groundwater discharge to Upper Klamath Lake in water-year quarter j of year k caused by a unit withdrawal at well i in water-year quarter j',

$qdr_{j,k,i,j'}$ is the reduction in groundwater discharge to drains in water-year quarter j of year k caused by a unit withdrawal at well i in water-year quarter j',

$rqs_{m,j,k,i,j'}$ is the response coefficient for groundwater discharge to stream reach m in water-year quarter j of year k caused by a unit withdrawal at well i in water-year quarter j',

$rql_{ukl,j,k,i,j'}$ is the response coefficient for groundwater discharge to Upper Klamath Lake in water-year quarter j of year k caused by a unit withdrawal at well i in water-year quarter j', and

$rqd_{j,k,i,j'}$ is the response coefficient for groundwater discharge to drains in water-year quarter j of year k caused by a unit withdrawal at well i in water-year quarter j'.

The response-matrix technique is based on the assumption that the numerical groundwater model is linear. In the case of a linear system, total drawdown and total reduction in groundwater discharge can be calculated using linear superposition

$$DD_{m,j,k} = \sum_{i=1}^{NW} \sum_{j'=3}^{4} rdd_{m,j,k,i,j'} Qw_{i,j'}. \tag{21}$$

$$QSR_{m,j,k} = \sum_{i=1}^{NW} \sum_{j'=3}^{4} rqs_{m,j,k,i,j'} Qw_{i,j'}. \tag{22}$$

$$QLR_{ukl,j,k} = \sum_{i=1}^{NW} \sum_{j'=3}^{4} rql_{ukl,j,k,i,j'} Qw_{i,j'}. \tag{23}$$

$$QDR_{j,k} = \sum_{i=1}^{NW} \sum_{j'=3}^{4} rqd_{j,k,i,j'} Qw_{i,j'}. \tag{24}$$

The response coefficients are the link between the simulation and optimization models of the upper Klamath Basin groundwater system. The response coefficients provide a compact version of the simulation model that calculates the effects of pumping on drawdown and groundwater discharge at constraint sites. The response coefficients are incorporated into the groundwater management model by substituting the right-hand sides of equations 21–24 into the left-hand sides of equations 8–14, to obtain the optimization model

$$maximize \sum_{i=1}^{NW} \sum_{j'=3}^{4} Qw_{i,j'}. \tag{25}$$

$$\sum_{i=1}^{NW} \sum_{j'=3}^{4} \left(rdd_{m,3,k,i,j'} - rdd_{m,2,k,i,j'} \right) Qw_{i,j'} \le DD_{seas,max}. \tag{26}$$

$$\sum_{i=1}^{NW} \sum_{j'=3}^{4} \left(rdd_{m,4,k,i,j'} - rdd_{m,2,k,i,j'} \right) Qw_{i,j'} \le DD_{seas,max}. \tag{27}$$

$$\sum_{i=1}^{NW} \sum_{j'=3}^{4} \left(rdd_{m,2,k+1,i,j'} - rdd_{m,2,k,i,j'} \right) Qw_{i,j} \le DD_{year,max}. \tag{28}$$

$$\sum_{i=1}^{NW} \sum_{j'=3}^{4} \left(rdd_{m,2,k+10,i,j'} - rdd_{m,2,k,i,j'} \right) Qw_{i,j'} \le DD_{10yr,max}. \tag{29}$$

$$\sum_{i=1}^{NW} \sum_{j'=3}^{4} rqs_{m,j,k,i,j'} Qw_{i,j'} \le QSR_{m,j,k,max}. \tag{30}$$

$$\sum_{i=1}^{NW} \sum_{j'=3}^{4} rql_{ukl,j,k,i,j'} Qw_{i,j'} \le QLR_{ukl,j,k,max}. \tag{31}$$

$$\sum_{i=1}^{NW} \sum_{j'=3}^{4} rqd_{j,k,i,j'} Qw_{i,j'} \le QDR_{j,k,max}. \tag{32}$$

$$\sum_{i=1}^{NW} Qw_{i,4} \le Qw_{4,min}. \tag{33}$$

$$0 \le Qw_i \le Qw_{i,max}. \tag{34}$$

Sequential Linear Programming

Complications can arise in the use of the response-matrix technique if the numerical model is nonlinear. The model of the upper Klamath Basin groundwater system has a number of recharge and discharge components that are simulated as piece-wise linear functions of the calculated hydraulic head. These boundary conditions can create nonlinear relations between hydraulic heads and discharges to or from a boundary. Examples of nonlinear relations important to the upper Klamath Basin model are those between hydraulic heads and groundwater discharge to streams, lakes, and drains found in the groundwater-discharge depletion constraints (equations 30–32). Because of these nonlinearities, the response coefficients can change as the simulated withdrawal conditions change. These types of nonlinearities have been handled in groundwater simulation-optimization problems by iterative methods that linearize the nonlinear head-dependent boundary equations (see, for example, Danskin and Gorelick, 1985; Danskin and Freckleton, 1989; Gorelick and others, 1993; Ahlfeld and Mulligan, 2000; Reichard and others, 2003; Ahlfeld and Baro-Montes, 2008).

Sequential linear programming was used in this study to address the nonlinearities associated with head-dependent boundary conditions. Sequential linear programming solves a series of linear programming subproblems, each formulated using the response matrix technique. In each iteration of the method, the response coefficients are calculated using a unit rate added to the current managed flow rates; the current managed rates are derived from the previous iteration's optimal withdrawal rates. The optimization problem (equations 25–34) is formulated and solved to provide updated managed withdrawal rates. The sequential process is continued until convergence is achieved. The convergence criterion used in this study is based on the objective function value (equation 25) and requires that the change in this value from iteration l to iteration $l +1$ be less than 0.05 percent. Two to eight iterations of the sequential linear programming method were required in the optimization results presented below. The model was formulated in each iteration of the sequential linear programming method using the response-matrix technique and was solved using the optimization software package LINDO (LINDO Systems, Inc., 2005).

Transient Model Linked with Optimization

The coupled simulation-optimization model was designed to simulate dynamic equilibrium conditions in the same manner previously described for the example simulations. Formulated in this way, there is no net change in storage over the annual hydrologic cycle. The baseline model is intended to simulate average annual hydrologic conditions during the 1970 to 2004 transient calibration period. The optimized model simulates the same conditions as the baseline model, with the addition of increased groundwater pumping in the Project area. The spatial and temporal distribution of differences in hydraulic head and groundwater discharge between the baseline and optimized models are the basis by which the numerical model of the upper Klamath Basin groundwater system is incorporated into the groundwater management model. In the baseline dynamic-equilibrium model, simulated hydraulic head and groundwater discharge vary from quarter to quarter, but return to their initial state at the end of the annual cycle. In the optimized dynamic-equilibrium model, simulated head and discharge depart from the baseline conditions and reach a new state of dynamic equilibrium that is a function of the groundwater withdrawal identified by the optimization model. The dynamic-equilibrium model approach ensures that withdrawal strategies identified by the optimization model could be sustained without causing perpetual reductions in groundwater storage and discharge. The simulation-optimization models presented in this report were simulated with a twenty-year model horizon to allow the optimized withdrawal to reach a new state of dynamic equilibrium.

Utilizing the dynamic-equilibrium model approach allows easy isolation of the impacts from supplemental pumping. The simulations do not, however, account for other external influences. Historical weather patterns in the upper Klamath Basin include decadal scale wet and dry cycles. Moreover, the background pumping in the example simulations does not include pumping historically associated with supplemental water rights in Oregon. These additional external influences, when incorporated into the simulation-optimization model, will affect the solution. Results presented in this report, therefore, are intended to demonstrate the utility of simulation-optimization model and should not be used for management purposes.

Evaluation of Selected Alternatives

Groundwater development in the upper Klamath Basin was evaluated with the aid of the coupled groundwater simulation-optimization model described in this report. Alternative management model formulations were examined to quantify relations between potential groundwater-withdrawal rates and limits on groundwater drawdown, depletions in groundwater discharge, and

groundwater demand. A varied set of management model applications were analyzed for alternative values of the drawdown, discharge, and demand constraints. Six sets of analyses were completed:

1. Base case. Drawdown limits defined as 20 ft for seasonal drawdown (equations 26 and 27), 4 ft for year-to-year drawdown (equation 28), and 25 ft for 10-year drawdown (equation 29); groundwater-discharge depletion constraints defined as 6 percent of baseline discharge for streams and lakes (equations 30 and 31); groundwater discharge depletion constraints defined as 20 percent of baseline discharge for drains (equation 32); no seasonal demand constraint.

2. Analysis of the sensitivity of model results to the upper bound on changes to groundwater discharge to drains (equation 32).

3. Analysis of the sensitivity of model results to the upper bound on seasonal drawdown (equations 26 and 27).

4. Analysis of the sensitivity of model results to the upper bound on year-to-year drawdown (equation 28).

5. Analysis of the sensitivity of the model results to the 4th quarter (July–September) withdrawal demand (equation 33).

6. Analysis of the sensitivity of the model results to the upper bound on changes to groundwater discharge to the Lost River.

Base-Case Results

The results of the example base-case simulation-optimization model are shown in figures 57–58. The optimization model provides 3rd and 4th quarter pumping rates for each of the managed wells shown in figure 55A. These represent the types of information, along with their spatial and temporal patterns, that are useful for practical implementation of a groundwater-development strategy. The solution of the sequential linear programming method gave a total withdrawal at managed wells of about 56,000 acre-ft during the April–September irrigation season, with quarterly rates of 198.3 ft^3/s during April–June (3rd quarter) and 112.9 ft^3/s during July–September (4th quarter). The managed pumping represents an increase of about 35 percent relative to pre-2001 pumping in the basin. Again, these figures do not reflect effects of drought cycles or nearby off-project supplemental pumping that will affect results. Figure 57 presents the optimal pumping by subregion and water-year quarter. Pumping by subregion ranges from approximately 1,000 acre-ft for the TID well group to approximately 29,000 acre-ft for the Klamath Valley well group. As a result of the drawdown and discharge restrictions imposed by the

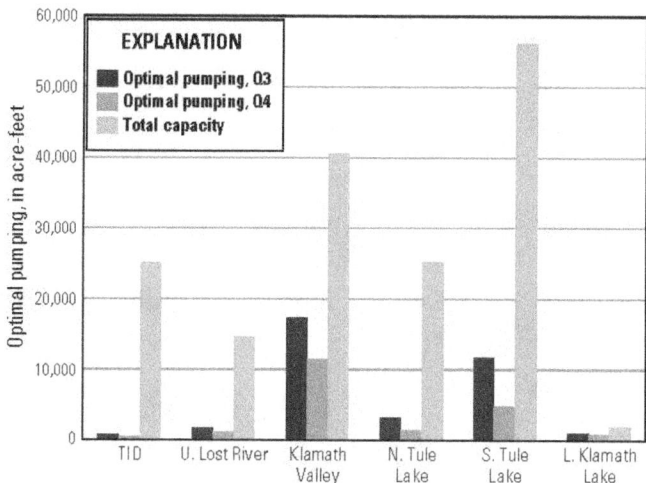

Figure 57. Summary of base-case optimization results for well groups. TID, Tulelake Irrigation District wells. U.Lost River, upper Lost River wells. N. Tule Lake, Northern Tule Lake subbasin wells. S. Tule Lake, Southern Tule Lake subbasin wells. L. Klamath Lake, Lower Klamath Lake wells. Q3, third quarter. Q4, fourth quarter.

optimization model, there is significant unused pumping capacity in all but the Lower Klamath Lake well group. The wells of the TID group (fig. 55A) contribute the smallest amount to total pumping when measured as a percentage of a group's pumping capacity (defined as the amount that could be withdrawn if all wells were pumped at their maximum rates). Of the total pumpage, 64 percent occurs as withdrawal during the 3rd quarter of the water year. By apportioning the majority of the pumping to the 3rd quarter, the optimization model is able to increase total pumping while meeting the year-to-year drawdown constraints. The ability of the groundwater system to accommodate increased 4th quarter pumping will be tested through sensitivity analysis.

Figures 58A and 58B show the distribution of pumping (geographically and with depth by model layer) for water-year quarters three and four respectively. The highest pumping rates are located in the southern Tule Lake and Klamath Valley areas. Eighty percent of the total pumping occurs at wells located in model layers 1 and 2 (denoted by TSY and TSO in table 1 and fig. 11); the remainder occurs at wells located in model layer 3 (denoted by TSV3 in table 1 and fig. 11). The difference in managed pumping with depth occurs despite there being approximately the same total capacity of the managed wells in layer 3 as there is in layers 1 and 2, and can be attributed in part to the lower storativity of the deeper zone of mixed sediments and volcanics.

The results presented in figures 57 and 58 show a concentration of pumping in the Klamath Valley and southern Tule Lake areas. Water managers may wish to consider

pumping patterns that redistribute some withdrawal to other areas. The optimization results can be analyzed to determine the trade-offs associated with increased (or decreased) pumping at selected wells. One measure of this trade-off is the reduced cost associated with each decision variable. Reduced cost is a local sensitivity that measures the change in the objective function resulting from a small change in the value of each pumping-rate decision variable (Gill and others, 1981; Ahlfeld and Mulligan, 2000). The reduced cost can be described alternatively as an 'increased benefit,' because if a pumping rate at a given location is at its upper bound, $Qw_{i,max}$, then the reduced cost is positive and represents the increase in the objective (equation 25) that would result from increasing $Qw_{i,max}$. In this case, the reduced cost measures the value of increased pumping capacity at location i. If Qw_i is greater than zero but less than $Qw_{i,max}$ the reduced cost is zero. If, however, Qw_i equals zero, the reduced cost is negative and represents the reduction in the objective (equation 25) that results from increasing Qw_i from zero. In this case, the reduced cost measures the penalty incurred when imposing a non-zero pumping rate at location i.

The reduced costs for the base-case optimization results are presented in figure 59. It can be seen that the greatest benefit would be obtained by increasing the capacity of a small number of wells found in the Klamath Valley, upper Lost River, and southern Tule Lake areas. All but one of the wells with reduced costs greater than 40 are located in model layers 1 and 2 (TSO and TSY in table 1 and fig. 11). Reduced costs that are negative indicate that increased pumping would be detrimental to the objective function. It can be seen in figure 59 that the greatest penalty would result from requiring pumping at wells with reduced costs less than –30 in the Klamath Valley, upper Lost River, northern Tule Lake, and southern Tule Lake areas. These wells are found in model layer 3 (TSV in table 1 and figure 11). There are also wells distributed throughout the Project area that have negative reduced costs between –15 and 0, indicating a relatively minor penalty if these wells are required to withdraw groundwater. These locations might be candidates for increased pumping if water managers required a different geographic distribution of pumping.

The locations of the constraints that limit withdrawal are shown in figure 60 and include 61 seasonal-drawdown constraints (38 in the 3rd quarter and 23 in the 4th quarter, figs. 60A and 60B), 12 year-to-year drawdown constraints (fig. 60C), groundwater discharge depletion constraints for two reaches of the Lost River (fig. 60A), and the drain discharge depletion constraint. All binding drawdown constraints occur in year 1 of the optimization model; the binding discharge constraints occur in year 20. The optimization results can be analyzed to identify the constraints that impose the greatest control on pumping. One way to do this is by examining the shadow price associated with each binding constraint.

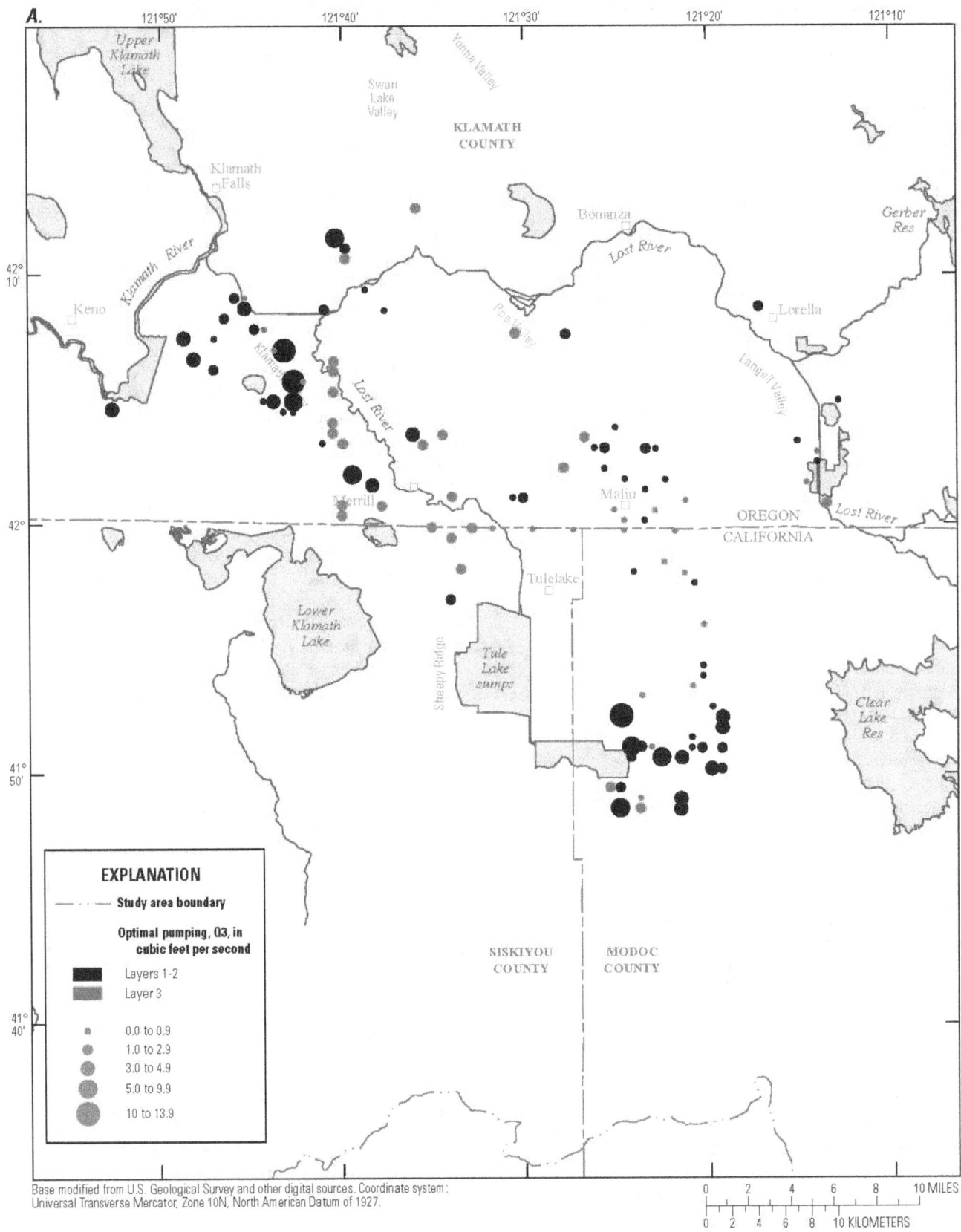

Figure 58. Optimal pumping rates for the base-case optimization model solution. (*A*) Optimal pumping rates in water-year quarter 3 (Q3). (*B*) Optimal pumping rates water-year quarter 4 (Q4).

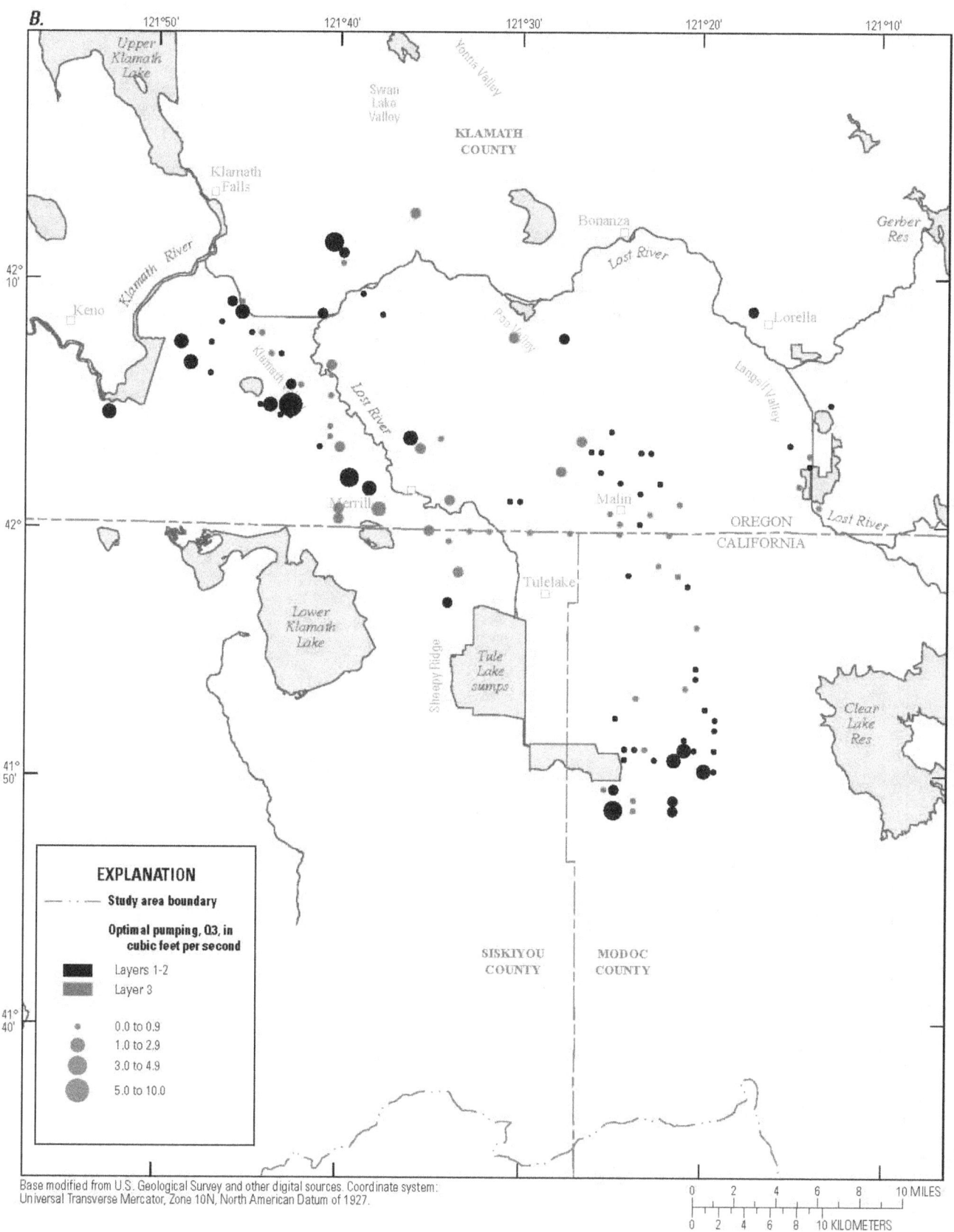

B.

EXPLANATION

— ·· — Study area boundary

Optimal pumping, Q3, in
cubic feet per second

■ Layers 1-2
■ Layer 3

· 0.0 to 0.9
● 1.0 to 2.9
● 3.0 to 4.9
● 5.0 to 10.0

Base modified from U.S. Geological Survey and other digital sources. Coordinate system:
Universal Transverse Mercator, Zone 10N, North American Datum of 1927.

0   2   4   6   8   10 MILES

0   2   4   6   8   10 KILOMETERS

Figure 58.—Continued

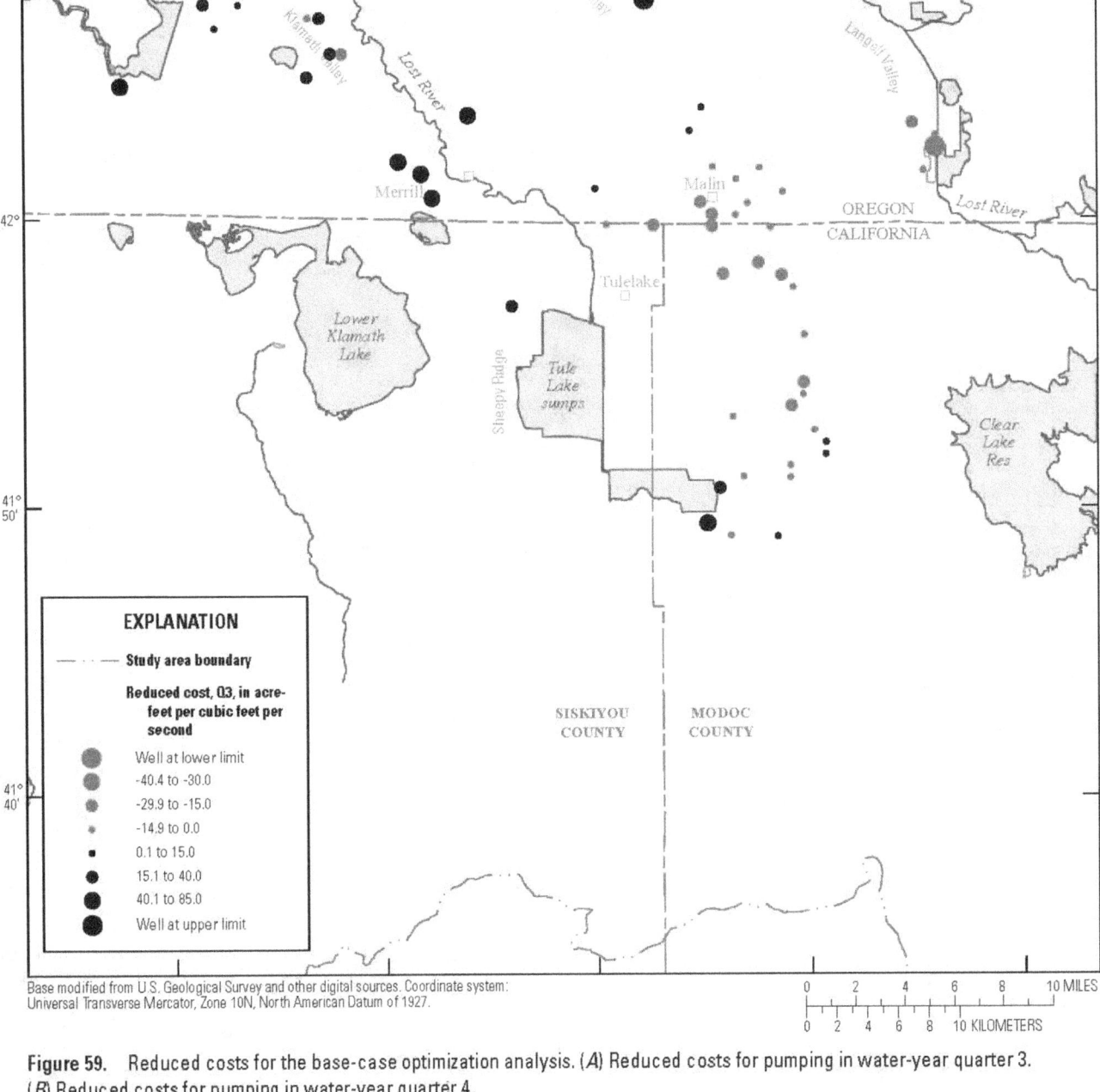

Figure 59. Reduced costs for the base-case optimization analysis. (*A*) Reduced costs for pumping in water-year quarter 3. (*B*) Reduced costs for pumping in water-year quarter 4.

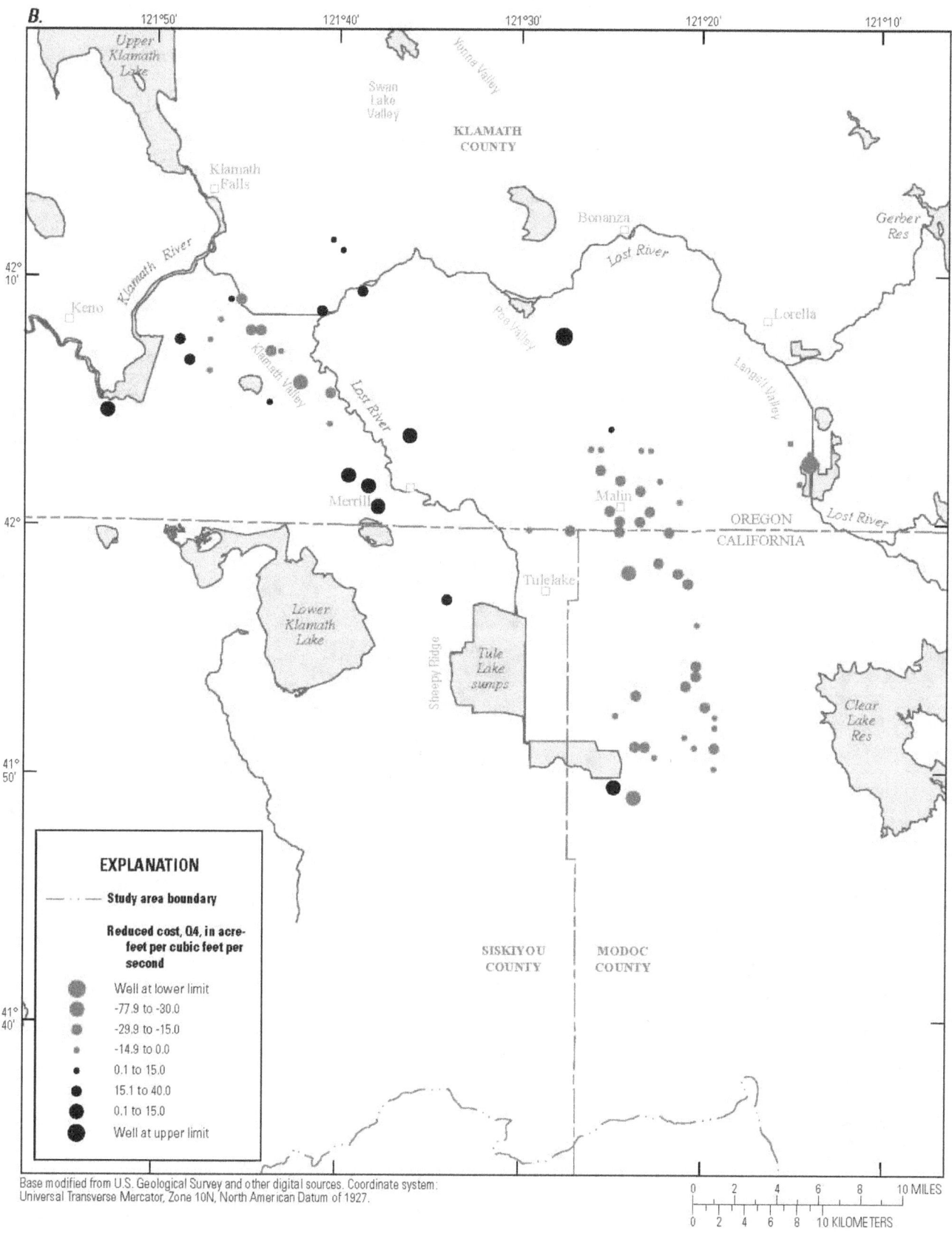

B.

EXPLANATION

— · · —   Study area boundary

Reduced cost, Q4, in acre-feet per cubic feet per second

- Well at lower limit
- -77.9 to -30.0
- -29.9 to -15.0
- -14.9 to 0.0
- 0.1 to 15.0
- 15.1 to 40.0
- 0.1 to 15.0
- Well at upper limit

Base modified from U.S. Geological Survey and other digital sources. Coordinate system:
Universal Transverse Mercator, Zone 10N, North American Datum of 1927.

0   2   4   6   8   10 MILES

0   2   4   6   8   10 KILOMETERS

Figure 59.—Continued

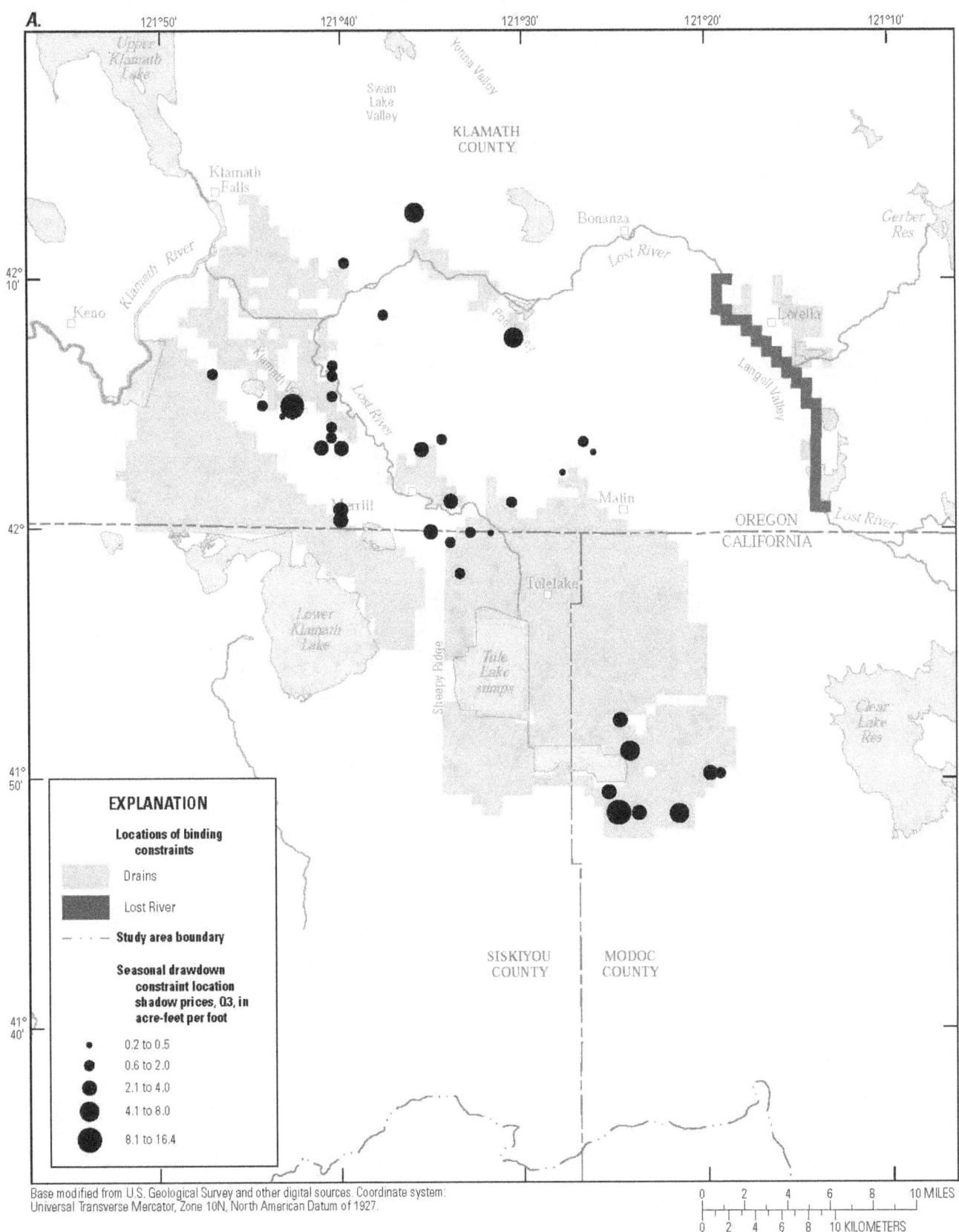

Figure 60.   Shadow prices for the base-case optimization analysis. (*A*) Shadow prices for binding seasonal drawdown constraints in water-year quarter 3. (*B*) Shadow prices for binding seasonal drawdown constraints in water-year quarter 4. (*C*) Shadow prices for binding year-to-year drawdown constraints.

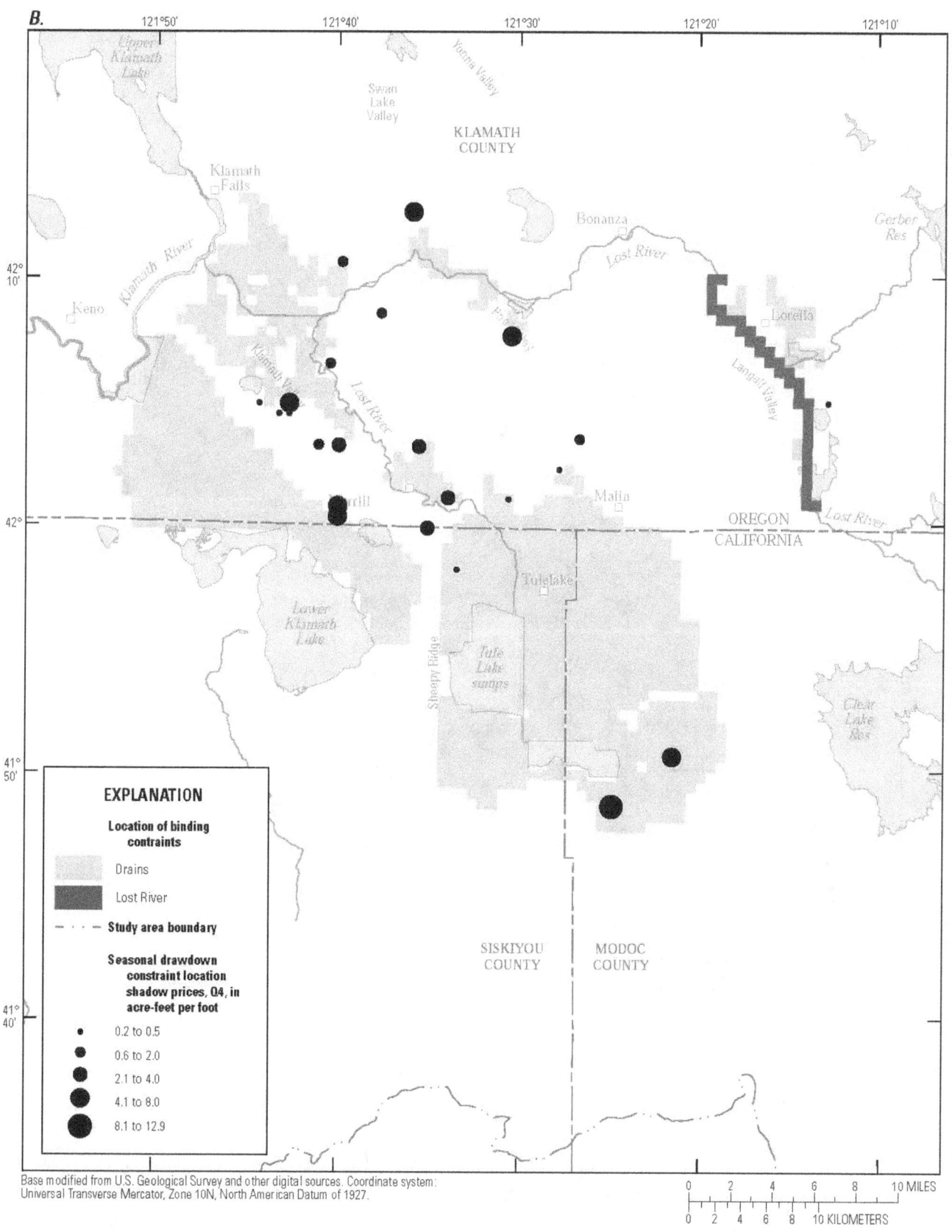

B.

EXPLANATION

Location of binding contraints

Drains

Lost River

— · · — · · — **Study area boundary**

**Seasonal drawdown
constraint location
shadow prices, Q4, in
acre-feet per foot**

•   0.2 to 0.5

●   0.6 to 2.0

●   2.1 to 4.0

●   4.1 to 8.0

●   8.1 to 12.9

Base modified from U.S. Geological Survey and other digital sources. Coordinate system:
Universal Transverse Mercator, Zone 10N, North American Datum of 1927.

0   2   4   6   8   10 MILES

0   2   4   6   8   10 KILOMETERS

Figure 60.—Continued

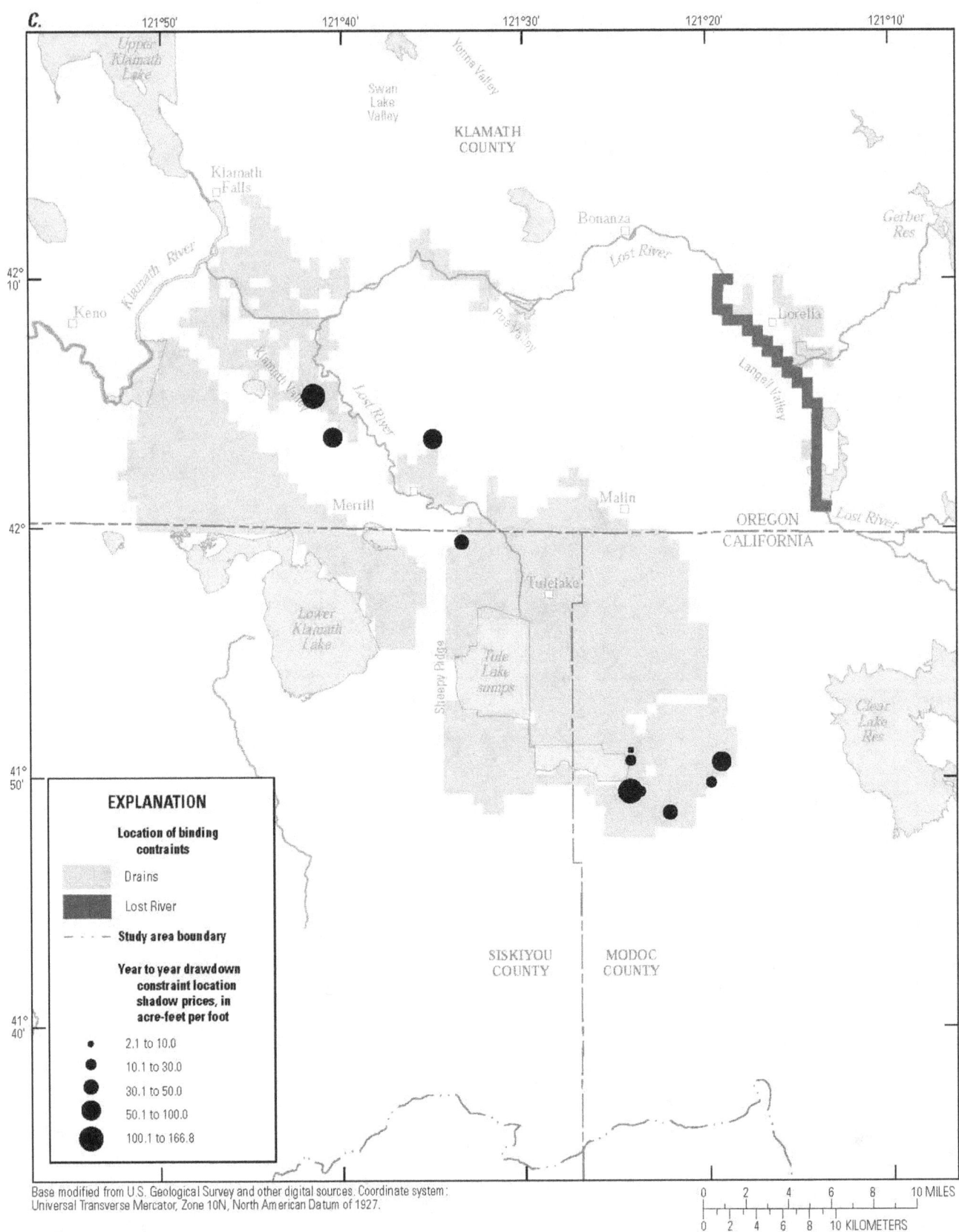

Figure 60.—Continued

The shadow price is a local sensitivity that measures the marginal utility of relaxing a constraint, and it is defined as the amount by which the objective would change if the value of that constraint were changed by one unit (Hillier and Lieberman, 1980; Ahlfeld and Mulligan, 2000). The shadow price can be used to evaluate the potential gain in total withdrawal that results if a binding constraint is relaxed. Alternatively, water managers may wish to impose more restrictive limitations on drawdown or groundwater discharge. In this case, the shadow price can be used to estimate the reduction in total withdrawal associated with tighter constraint limits.

The shadow prices for the binding seasonal-drawdown constraints range from about 0.2 to about 16 with an average of 2.7 (figs. 60A and 60B). The largest shadow prices for seasonal drawdown constraints are found in the southern Tule Lake and Klamath Valley areas. The magnitude of the shadow price for seasonal-drawdown constraints suggests there would be little improvement of the optimization objective if an individual constraint were relaxed. The results also indicate there would be little reduction in total pumping if a single constraint were tightened to impose a more restrictive drawdown limit. The optimization objective shows greater sensitivity to the binding year-to-year drawdown constraints and the binding discharge constraints. Shadow prices for the binding year-to-year drawdown constraints (fig. 60C) range from about 2 to about 170 (the average across all binding year-to-year drawdown constraints is about 54), with the largest shadow prices found in the southern Tule Lake area. The shadow price for the binding discharge-to-drains constraint is about 1,050, and the average shadow price for the binding Lost River-discharge constraints is about 300. The shadow price values indicate that the greatest increase in total withdrawal would result from relaxing $QDR_{j,k,max}$ the upper bound on allowable depletion in groundwater discharge to drains, and the smallest decrease in total withdrawal would result from tightening the seasonal-drawdown constraint, $DD_{seas,max}$. It should be noted that the large number of binding seasonal drawdown constraints could result in a significant increase (decrease) in total pumping if the seasonal drawdown limit, $DD_{seas,max}$, is uniformly increased (decreased) across all constraint locations. The sensitivities to drain discharge, seasonal drawdown, year-to-year drawdown, seasonal water demand, and Lost River-discharge constraint limits are evaluated the following sections.

It is important to note that the groundwater-discharge depletion constraints related to environmental flows in streams and lakes (equations 30 and 31) were not limiting. As described earlier in this report, the impact of pumping will depend on the rate of pumping, the hydraulic properties of the aquifer, and the proximity to hydrologic boundaries,

such as streams, lakes, drains, and ET surfaces. The managed wells are found in an area with an extensive network of agricultural drains, shallow groundwater (with associated evapotranspiration losses), Lower Klamath Lake, the Tule Lake sumps, and the Lost River. The water pumped by the managed wells comes primarily from these boundaries. The largest impact of the increased pumping to environmentally sensitive areas was determined to be along a reach of the Klamath River between Keno and John C. Boyle Reservoirs. The simulated effect of the optimized pumping is to decrease groundwater discharge to this reach by about 0.13 percent, which is well within the 6-percent limit defined in the KBRA and required by the optimization model formulation.

Vary Limit of Groundwater Discharge Constraints for Drains

The shadow prices presented in figures 60A–60C measure the local sensitivity of the objective to changes in the value of a binding constraint's right hand side. It is useful to test the impact of larger changes in constraint bounds on the objective value. The first set of analyses evaluates the sensitivities of the optimal groundwater withdrawals to changes in the constraint controlling groundwater discharge to drains (equation 32), which was set to 20 percent of the baseline discharge in the base-case analysis. The allowable reduction in groundwater discharge to drains was varied from 10 percent to 40 percent of baseline conditions. As shown in figure 61 and table 4, total groundwater withdrawal varies from about 33,000 acre-ft for a discharge-depletion limit of 10 percent to about 77,000 acre-ft for a limit of 40 percent and higher. As the drain-discharge depletion constraint is increased from the 10-percent limit, the total withdrawal increases in an approximately linear manner until the constraint limit reaches 30 percent. In other words, the marginal utility of increasing $QDR_{j,k,max}$ is approximately the same for values of $QDR_{j,k,max}$ as much as 30 percent. At a constraint limit of 40 percent and beyond, however, the discharge depletion constraint for drains is no longer binding and there is no utility in further relaxing this constraint, as indicated by the shadow price of 0 (fig. 61).

The tradeoff curve presented in figure 61 highlights the importance of understanding the spatial and temporal patterns of drain-water discharge and groundwater pumping within the Project. The optimization model indicates the total withdrawal could vary by approximately 44,000 acre-ft depending on the limit imposed on groundwater discharge to drains (with all other constraints fixed at their base-case limits). Improved information about the patterns of drain-water discharge and demand would reduce the uncertainty associated with the value of this constraint and would provide a better understanding of the potential for groundwater development.

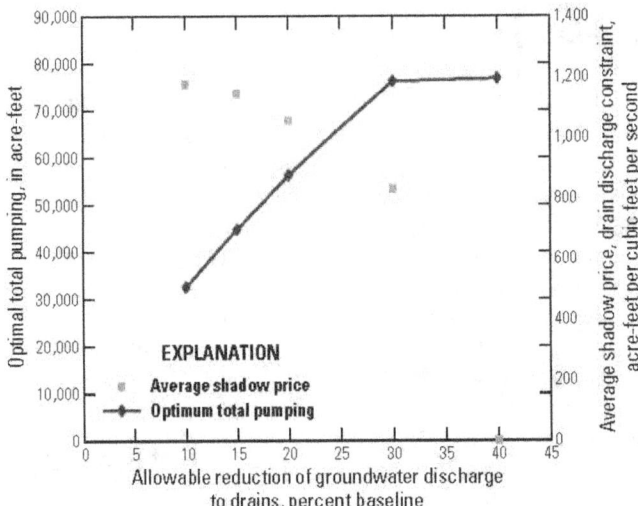

Figure 61. Sensitivity of optimization results to changes in the groundwater-discharge depletion constraint limit for drains (equation 32).

Table 4. Sensitivity of optimization model results to changes in groundwater discharge depletion constraint limit for drains (equation 32).

[All volumes are in acre-feet. **Abbreviations:** TID, Tulelake Irrigation District wells; KV, Klamath Valley wells; ULR, upper Lost River wells; NTL, Northern Tule Lake subbasin wells; STL, Southern Tule Lake subbasin wells; LKL, Lower Klamath Lake wells; Q3, third quarter; Q4, fourth quarter]

Well group and water-year quarter	Allowable reduction in groundwater discharge to drains, in percent				
	10	15	20 (base)	30	40
TID, Q3	0	542	705	1,916	1,916
TID, Q4	0	108	470	1,446	1,446
ULR, Q3	1,753	1,753	1,699	1,464	1,428
ULR, Q4	1,265	1,247	1,229	1,085	1,121
KV, Q3	11,911	15,834	17,352	17,985	17,424
KV, Q4	5,567	10,212	11,460	13,773	14,677
NTL, Q3	705	1,681	3,290	7,447	7,447
NTL, Q4	651	1,229	1,518	6,055	6,127
STL, Q3	5,766	6,688	11,731	16,322	15,273
STL, Q4	2,892	3,507	4,808	6,615	7,881
LKL, Q3	1,085	1,085	1,085	1,085	1,085
LKL, Q4	940	940	940	940	940
Total	32,535	44,826	56,286	76,132	76,765

Vary Limit of Seasonal Drawdown Constraints

The second set of analyses was developed to evaluate the sensitivities of the optimal groundwater withdrawals to changes in the seasonal-drawdown limit, which is designed to prevent excessive declines in groundwater levels over the irrigation season. In the base-case analysis, the seasonal-drawdown constraints had the smallest shadow prices, indicating the model solution was least sensitive to these limits. The seasonal-drawdown limit found in equations 26 and 27, $DD_{seas,max}$, was varied from 10 ft to 30 ft. The results are presented in figure 62 and table 5. Total groundwater withdrawal varied from about 52,000 acre-ft with the seasonal-drawdown limit of 10 ft, to about 57,000 acre-ft for a seasonal drawdown limit of 30 ft. As shown in figure 62, an increase in $DD_{seas,max}$ from 20 ft to 25 ft increases total pumping by about 1 percent, which is consistent with the small shadow price associated with binding

seasonal-drawdown constraints in the base-case solution (figs. 60A and 60B). Likewise, a reduction in $DD_{seas,max}$ from 20 ft to 15 ft has little impact on total withdrawal. However, the effect of reducing the seasonal drawdown limit shows a nonlinear pattern. Reducing the limit from 20 ft to 15 ft causes a reduction in total withdrawal of about 1,000 acre-ft. If the seasonal-drawdown constraint is further reduced to 10 ft, then the total withdrawal is reduced by an additional 3,000 acre-ft. The nonlinear pattern of this tradeoff curve is also reflected in the shadow prices, which increase in magnitude as the seasonal-drawdown constraint decreases. As the seasonal-drawdown constraint is tightened, there is also a shift in the type and location of binding constraints. For a seasonal-drawdown limit of 10 ft, the number of binding seasonal-drawdown constraints increases to 147 and the year-to-year drawdown constraints no longer influence the solution.

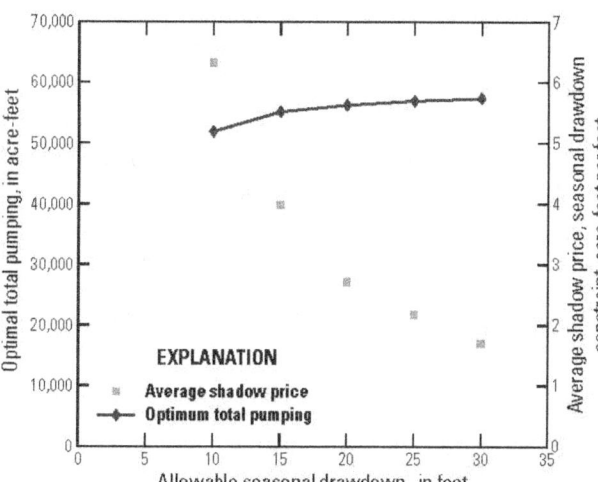

Figure 62. Sensitivity of optimization results to changes in the seasonal drawdown limit (equations 26–27).

Table 5. Sensitivity of optimization model results to changes in seasonal drawdown limit (equations 26–27).

[All volumes are in acre-feet. **Abbreviations:** TID, Tulelake Irrigation District wells; KV, Klamath Valley wells; ULR, upper Lost River wells; NTL, Northern Tule Lake subbasin wells; STL, Southern Tule Lake subbasin wells; LKL, Lower Klamath Lake wells; Q3, third quarter; Q4, fourth quarter]

Well group and water-year quarter	Seasonal drawdown limit, in feet				
	10	15	20 (base)	25	30
TID, Q3	976	831	705	886	1,066
TID, Q4	434	398	470	271	18
ULR, Q3	1,283	1,500	1,699	1,880	2,042
ULR, Q4	904	1,066	1,229	1,356	1,482
KV, Q3	13,647	16,123	17,352	18,220	18,364
KV, Q4	10,972	11,695	11,460	11,044	10,230
NTL, Q3	5,151	3,615	3,290	3,181	3,181
NTL, Q4	3,037	2,693	1,518	1,518	1,319
STL, Q3	9,164	10,899	11,731	12,381	12,869
STL, Q4	4,898	4,537	4,808	4,067	4,609
LKL, Q3	759	940	1,085	1,211	1,338
LKL, Q4	651	850	940	904	831
Total	51,875	55,147	56,286	56,918	57,352

Vary Limits of Year-to-Year Drawdown Constraints

The third sensitivity analysis tests the impact of changing the year-to-year drawdown limit on the optimization results. The year-to-year drawdown constraint (equation 28) limits the residual drawdown at the beginning of an irrigation season due to pumping that occurred in previous years. In the base-case optimization model, the year-to-year drawdown constraint was set to 4 ft. The optimal solution identified 12 locations with binding year-to-year drawdown constraints, with an average shadow price of 54 (fig. 60C). In this analysis, the year-to-year drawdown limit was varied from 2 ft to 8 ft. The results are presented in figure 63 and table 6. The results reveal a mildly nonlinear relation between total withdrawal and the year-to-year drawdown limit. Total withdrawal varies from about 53,000 acre-ft for a year-to-year drawdown limit of 2 ft to about 57,000 acre-ft for a limit of 8 ft.

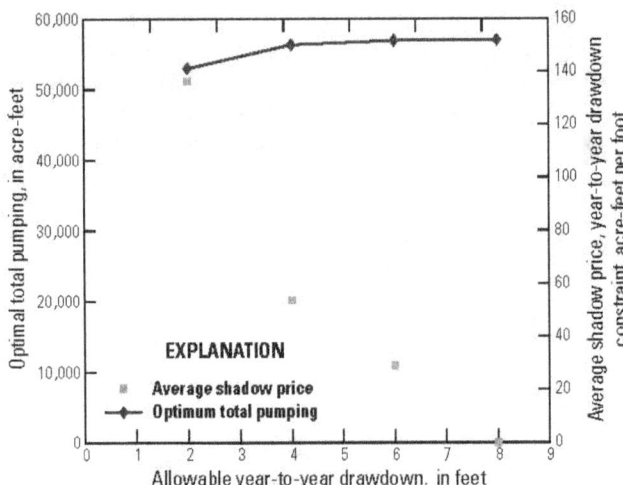

Figure 63. Sensitivity of optimization results to changes in the year-to-year drawdown constraint limit (equation 28).

Table 6. Sensitivity of optimization model results to changes in the year-to-year drawdown limit (equation 28).

[All volumes are in acre-feet. **Abbreviations:** TID, Tulelake Irrigation District wells; KV, Klamath Valley wells; ULR, upper Lost River wells; NTL, Northern Tule Lake subbasin wells; STL, Southern Tule Lake subbasin wells; LKL, Lower Klamath Lake wells; Q3, third quarter; Q4, fourth quarter]

Well group and water-year quarter	Year-to-year drawdown limit, in feet			
	2	4 (base)	6	8
TID, Q3	1,645	705	940	940
TID, Q4	488	470	560	560
ULR, Q3	1,554	1,699	1,717	1,717
ULR, Q4	1,085	1,229	1,229	1,229
KV, Q3	16,177	17,352	16,719	16,701
KV, Q4	8,333	11,460	11,821	11,839
NTL, Q3	6,778	3,290	2,241	1,934
NTL, Q4	2,296	1,518	1,265	1,265
STL, Q3	10,212	11,731	12,183	12,363
STL, Q4	3,037	4,808	6,109	6,254
LKL, Q3	705	1,085	1,085	1,085
LKL, Q4	560	940	940	940
Total	52,869	56,286	56,810	56,828

Vary Limits of Constraint for Seasonal Water Demand

The fourth sensitivity analysis tests the impact of the seasonal-demand constraint (equation 33). This constraint requires a minimum withdrawal rate from all managed wells during the 4th quarter of the water year. The base-case optimization model identifies 56,000 acre-ft of withdrawal, with about 20,000 acre-ft (36 percent) pumped in the 4th quarter of the water year (figs. 57 and 58B). In the fourth set of analyses, the minimum total withdrawal from all wells in the 4th quarter was systematically increased from approximately 23,000 acre-ft to approximately 45,000 acre-ft in increments of approximately 4,500 acre-ft (fig. 64 and table 7). For a 4th quarter demand of 23,000 acre-ft, total withdrawal decreased by about 100 acre-ft, and for a demand of 41,000 acre-ft, total withdrawal decreased by about 7,000 acre-ft. When demand was increased to 45,000 acre-ft, the model was infeasible because it was unable to identify a distribution of 4th quarter withdrawal rates that total 45,000 acre-ft while simultaneously limiting seasonal drawdown to 20 ft and year-to-year drawdown to 4 ft.

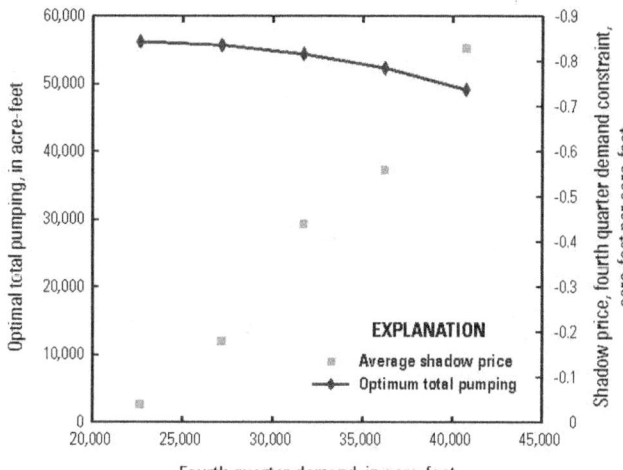

Figure 64. Sensitivity of optimization results to changes in the seasonal water demand limit (equation 33).

Table 7. Sensitivity of optimization results to changes in the seasonal water demand limit (equation 33).

[All volumes are in acre-feet. **Abbreviations:** TID, Tulelake Irrigation District wells; KV, Klamath Valley wells; ULR, upper Lost River wells; NTL, Northern Tule Lake subbasin wells; STL, Southern Tule Lake subbasin wells; LKL, Lower Klamath Lake wells; Q3, third quarter; Q4, fourth quarter]

Well group and water-year quarter	Minimum fourth quarter pumping, in cubic feet per second					
	0 (base)	125	150	175	200	225
TID, Q3	705	705	0	0	0	0
TID, Q4	470	470	687	1,012	1,392	958
ULR, Q3	1,699	1,681	1,627	1,428	1,374	723
ULR, Q4	1,229	1,211	1265	1,374	1,392	1,500
KV, Q3	17,352	17,207	15,472	11,857	7,718	3,272
KV, Q4	11,460	11,586	13,249	14,912	16,792	17,099
NTL, Q3	3,290	2,494	1,247	669	217	0
NTL, Q4	1,518	3,272	4,211	5,007	5,368	5,549
STL, Q3	11,731	10,429	9,182	7,772	5,874	4,121
STL, Q4	4,808	5,097	6,742	8,405	10,267	14,496
LKL, Q3	1,085	1,085	1,085	1,085	1,012	362
LKL, Q4	940	940	940	940	958	1,066
Total	56,286	56,177	55,707	54,460	52,363	49,146

Vary Limit of Groundwater Discharge Constraints for the Lost River

The Lost River is a source of irrigation water in the upper Lost River subbasin. The analyses presented thus far impose a 6-percent limit on the depletion in groundwater discharge to the Lost River. This sensitivity analysis evaluates the impact of increasing this limit to allow a greater reduction in groundwater discharge to the Lost River. The allowable reduction in groundwater discharge to the Lost River was varied from the base-case limit of 6 percent to 50 percent (fig. 65 and table 8). When the limit is increased to 15 percent, total pumping is about 58,000 acre-ft, which is approximately a 2-percent increase from the base case. For a 50-percent limit, total withdrawal is about 60,000 acre-ft.

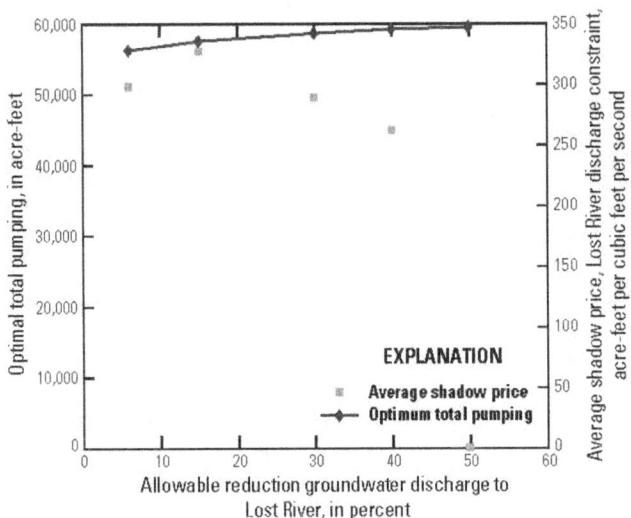

Figure 65. Sensitivity of optimization results to changes in the groundwater-discharge depletion limit for the Lost River, Oregon and California (equation 30).

Table 8. Sensitivity of optimization model results to changes in the groundwater discharge depletion limits for the Lost River, Oregon and California (equation 30).

[All volumes are in acre-feet. **Abbreviations:** TID, Tulelake Irrigation District wells; KV, Klamath Valley wells; ULR, upper Lost River wells; NTL, Northern Tule Lake subbasin; STL, Southern Tule Lake subbasin wells; LKL, Lower Klamath Lake wells; Q3, third quarter; Q4, fourth quarter]

Well group and water-year quarter	Allowable reduction in discharge to Lost River, in percent				
	6 (base)	15	30	40	50
TID, Q3	705	705	705	705	705
TID, Q4	470	380	380	398	398
ULR, Q3	1,699	2,874	3,778	4,049	4,049
ULR, Q4	1,229	2,006	2,765	3,326	3,778
KV, Q3	17,352	16,864	16,719	16,502	16,213
KV, Q4	11,460	11,460	11,441	11,333	10,953
NTL, Q3	3,290	3,037	3,579	3,073	3,434
NTL, Q4	1,518	1,410	1,573	1,464	1,428
STL, Q3	11,731	11,984	11,387	11,604	12,074
STL, Q4	4,808	4,808	4,338	4,772	4,483
LKL, Q3	1,085	1,085	1,085	1,085	1,085
LKL, Q4	940	940	940	940	940
Total	56,286	57,551	58,690	59,250	59,539

Model Limitations

Groundwater flow models are necessarily simplified mathematical representations of complex natural systems. Because of this, there are limits to the accuracy with which groundwater systems can be simulated. These limitations must be known when using models and interpreting model results.

There are many sources of error and uncertainty in models. Model error commonly stems from practical limitations of grid spacing, time discretization, parameter structure, insufficient calibration data, and the effects of processes not simulated by the model. These factors, along with unavoidable error in observations, result in uncertainty in model predictions.

Specific sources of uncertainty in the upper Klamath Basin regional model include grid spacing and parameter structure. The 2,500 ft by 2,500 ft grid spacing of the upper Klamath Basin regional groundwater model limits its ability to simulate conditions on smaller spatial scales. For example, because heads are averaged over areas of roughly a quarter square mile, drawdown in response to pumping wells at distances smaller than about 2,500 ft cannot easily be simulated. Because of the vertical discretizaton, conditions such as head changes due to pumping or other stresses are similarly averaged over vertical distances, limiting the ability to simulate effects to specific strata. Because of the limited availability of subsurface geologic information, hydraulic conductivity is simulated as uniform over broad areas, as shown in figure 6, and does not reflect the true complexity of the geology. Other parameters, such as streambed and lakebed conductance, are also simplified because of the lack of information.

The formulation of streams in a manner that only simulates groundwater discharge to streams and not stream leakage to the aquifer system is another limitation and potential source of uncertainty. In general, groundwater/surface-water interaction in the upper Klamath Basin is overwhelmingly dominated by groundwater discharge to streams. Seepage run data indicate leakage from streams to the aquifer system does not occur to a measureable degree along any of the major streams, and stream leakage is not a significant source of recharge. Should simulated head changes result in groundwater levels dropping below stream elevations in normally-gaining reaches, the model would not simulate the possible addition of water to the groundwater system from stream leakage. Should this occur, it is not likely to affect simulation results on a regional scale, but could affect simulated conditions near the affected stream reach.

Model error and uncertainty are not uniformly distributed. The model fit to observations is best where there are abundant data. Simulated conditions are more uncertain where data are sparse, such as unpopulated upland areas.

The upper Klamath Basin regional groundwater model was intended to simulate groundwater flow over an 8,000 mi² area. Groundwater management issues and specific questions continued to evolve after the model was constructed. Therefore, the model is not necessarily optimized to address all current groundwater management questions. As demonstrated in previous sections of this report, however, the model does a good job of simulating the spatial distribution of hydraulic head throughout the basin as well as the distribution of groundwater discharge to the stream network. The model also does a reasonable job of simulating the broad response of the groundwater system to climate influences such as decadal drought cycles and longer term trends, and is able to simulate the effects of large-scale irrigation pumping as observed in and around the Bureau of Reclamation's Klamath Project area. Therefore, even with the limitations described above, the model can be a useful tool for informing groundwater management in the basin.

The numerical results of the flow model have an associated, but un-quantified, uncertainty. While it is possible to quantify model prediction uncertainty, that analysis is not included within the scope of this report. A sense of model uncertainty will develop as conditions are monitored in the future and compared to model predictions. For these reasons, continued monitoring of hydrologic conditions in the basin is crucial. For practical purposes it is advisable to maintain an adaptive approach whereby management strategies can shift if observations differ from model predictions. Water managers should also be on the lookout for local anomalies resulting from geologic complexity not represented in the model. Model error and uncertainty can be reduced in the future by further model refinements and collection of new calibration data.

Next Steps

The development of groundwater flow and management models presented in this report is an important step in understanding the regional groundwater system in the upper Klamath Basin and how that system can be managed. The flow model can be used alone to evaluate the response of the groundwater system to any variety of future pumping, water management, or climate conditions. When used with a groundwater management model, as demonstrated in this report, the coupled models can be used to evaluate optimal strategies for meeting water management objectives while honoring predefined limits on impacts.

The flow model simulates the regional distribution of hydraulic head and groundwater discharge to streams, as well as the climate-driven fluctuations in water levels and groundwater discharge. It also simulates the water-level response to large-scale irrigation pumping. Certain refinements to the model, such as finer vertical discretization and improved representation of critical spring areas, could help fine-tune the model for addressing current management objectives.

A logical next step with simulation-optimization modeling will be to work with resource management agencies, water users, and other stakeholders to refine the management model to incorporate a more complete set of groundwater management objectives and constraints consistent with the full set of regulatory limits and practical (operational) considerations. When used in actual application, the simulation-optimization model must include realistic climate variability and background (off project) supplemental pumping rates.

Summary

The permeable volcanic bedrock of the upper Klamath Basin hosts a substantial regional groundwater flow system that provides much of the flow to major streams and lakes. These streams and lakes, in turn, provide water for wildlife habitat and are the principal source of irrigation water for the basin's agricultural economy. Increased allocation of surface water for aquatic wildlife in the past decade has resulted in increased reliance on groundwater for irrigation. The potential effects of increased groundwater pumping on groundwater levels and discharge to springs and streams has caused concern among irrigators dependent on groundwater, resource managers, wildlife biologists, and other stakeholders. In order to better understand the groundwater hydrology of the basin, to provide information on the potential impacts of increased groundwater development, and to aid in the development of groundwater management strategies, the U.S. Geological Survey (USGS), in collaboration with the Oregon Water Resources Department and the U.S. Bureau of Reclamation, developed a groundwater flow model that can simulate the response of the hydrologic system to these new stresses.

The flow model, which is described in this report, was developed using the USGS MODFLOW finite-difference modeling code. Model cells have lateral dimensions of 2,500 feet (ft) by 2,500 ft and are aligned in a grid consisting of 285 east-west trending rows and 210 north-south trending columns covering the entire upper Klamath Basin. In the vertical dimension, the model consists of three layers of varying thicknesses ranging from about 5 ft to 3,600 ft, depending on topography and proximity to the edge of the model. Hydraulic characteristics of subsurface materials are represented in 18 hydraulic parameter zones reflecting large-scale geologic conditions. Hydraulic parameter zonation is simpler at depth due to the lack of detailed geologic information.

Boundary conditions include specified-flux boundaries and head-dependent flux boundaries. Most boundaries with adjacent basins, as well as the contact with underlying low-permeability early Tertiary strata, are formulated with specified fluxes of zero. Groundwater recharge and pumping are simulated as specified fluxes varying each quarterly stress period. Head-dependent flux boundaries include streams, lakes and reservoirs, agricultural drains, evapotranspiration directly from aquifers in areas of shallow groundwater, and boundaries with adjacent basins in selected areas. All major streams and most major tributaries with substantial groundwater discharge are included in the model.

The model was calibrated using inverse methods to transient conditions from 1989 to 2004. Calibration data included 5,636 head measurements from 663 wells. Of these, 444 wells had time series consisting of 2 to 64 observations. Estimates of average groundwater discharge were available for 52 stream reaches or spring complexes. Time series of estimated groundwater discharge were available for 10 stream reaches or springs. The calibration data show that the groundwater system in the upper Klamath Basin responds to decadal climate cycles, with groundwater levels and spring flows rising and falling in response to wet and dry periods. Groundwater levels also show seasonal and year-to-year fluctuations in response to groundwater pumping.

Calibrated hydraulic conductivity values span nearly four orders of magnitude, ranging from 5.9×10^{-6} feet per second (ft/s) for Quaternary volcanic rocks in the southern part of the model area to 1.2×10^{-2} ft/s for Mazama tephra deposits. Late Tertiary volcanic deposits range from 1.0×10^{-5} ft/s to 9.3×10^{-4} ft/s. Quaternary volcanic deposits (other than Mazama tephra deposits) range from 5.9×10^{-6} ft/s to 4.0×10^{-5} ft/s. Late Tertiary sedimentary strata range from 2.9×10^{-4} ft/s to 3.5×10^{-3} ft/s, and the calibrated hydraulic conductivity for Quaternary sediments is 5.8×10^{-3} ft/s. Calibrated specific storage values range from 7.5×10^{-7} ft^{-1} to 1.0×10^{-3} ft^{-1}, with the smallest values more common with increasing depth. Vertical anisotropy (the ratio of horizontal hydraulic conductivity to vertical hydraulic conductivity) ranges from 10 to 1,000.

Model fit is evaluated by looking at the magnitude and distribution of the differences between field observations of heads and fluxes and their simulated equivalents (known as residuals). Fitted error statistics indicate that simulated heads are on average within about 30 ft of field measurements. Head residuals, which should ideally be random, show some geographical clustering. This is probably an artifact caused by the lack of detailed subsurface geologic information in many areas, and by the representation of spatially variable hydraulic properties in broad, uniform zones. Heads show a slight negative bias, meaning that simulated values have a tendency to be higher rather than lower compared to measurements. This is likely an artifact of the coarse vertical discretization. Visual comparisons of time series of simulated and measured heads show that the model simulates observed climate-driven water-level fluctuations over most of the model area. The model also simulates pumping-caused water-level changes in heavily pumped areas around the Klamath Reclamation Project. Visual comparison of time series of simulated and measured groundwater discharge to stream reaches shows that the model captures both the overall volumes and climate-driven fluctuations of groundwater discharge to major streams. Pumping effects are generally not visually detectable in streamflow or groundwater-discharge records.

The model has the ability to simulate the effects of external stresses, such as pumping or climate variations, on the water levels and groundwater discharge to streams, lakes, drains, and other boundaries. Example model simulations show that the timing and location of the effects of groundwater pumping vary markedly depending on the pumping location. Pumping from wells close (within a few miles) to groundwater-discharge features, such as springs, drains, and certain streams, can affect those features within weeks or months of the onset of pumping, and the impacts can be essentially fully manifest in several years. However, simulations indicate that responses to seasonal variations in pumping rates are buffered by the groundwater system, and peak impacts are closer to mean annual pumping rates than to instantaneous pumping rates. In other words, pumping effects are spread out over the entire year. When pumping locations are distant (more than several miles) from discharge features, the effects take many years or decades to fully impact those features, and much of the pumped water comes from groundwater storage over a broad geographic area even after two decades. Moreover, because the effects are spread out over a broad area, the impacts to individual features are much smaller than in the case of nearby pumping. Simulations show that the discharge features most affected by pumping in the area of the Klamath Reclamation Project are agricultural drains, and impacts to other surface-water features are small in comparison. Reductions in discharge to agricultural drains could potentially have operational considerations for Reclamation Project managers; reductions could also have ramifications with regard to refuge water supplies.

Developing a groundwater management strategy in the upper Klamath Basin requires understanding the effects of a wide range of possible pumping scenarios on groundwater levels and discharge, and identifying the best pumping strategy to meet water-user needs while not resulting in unacceptable impacts. To meet this need, a groundwater management model was developed that uses techniques of constrained optimization along with the groundwater flow model to identify the optimal strategy to meet water-user needs while honoring defined constraints on impacts to groundwater levels or streams. The coupled models are referred to as groundwater simulation-optimization models.

Example groundwater simulation-optimization models were formulated to demonstrate their utility in developing strategies to meet water demand in the upper Klamath Basin. The models maximize groundwater pumping while simultaneously avoiding the detrimental impacts of pumping on groundwater levels and discharge. Total groundwater withdrawals were calculated under alternative constraints for drawdown, reductions in groundwater discharge to surface water, and for water demand to understand the potential benefits and limitations for groundwater development in the upper Klamath Basin.

The initial application of the simulation-optimization model was made with the base-case constraint definitions that limit seasonal, year-to-year, and long-term drawdowns, limit reductions in groundwater discharge to selected streams, and limit reduction in groundwater discharge to the Klamath Project drain system. Given the example constraints and current well configuration, the optimization analysis identified approximately 56,000 acre-ft per year of groundwater that can be pumped on an annual basis in addition to the background pumping fixed at the 2000 pumping rate, with 64 percent of the total pumping occurring in the third quarter of the water year and 80 percent occurring in model layers 1 and 2. Subsequent model applications indicated that changes in the groundwater-discharge, drawdown, and water-demand constraint limits could result in substantial changes in optimal allowable groundwater withdrawal. It is important to note that the demonstration exercise does not include historic climate variability or off-project (but nearby) supplemental irrigation pumping, both of which will affect results.

The sensitivity of the optimal solution to the model constraints was tested by modifying their limits. The sensitivity of the solution to the drain-discharge constraints was tested by varying the upper bound on the allowable reduction in groundwater discharge to the drain system. Total withdrawal calculated by the optimization model ranged from approximately 33,000 acre-ft for a 10-percent constraint to approximately 77,000 acre-ft for a 40-percent limit. The sensitivity of the solution to the seasonal and year-to-year drawdown constraints' limits was also tested. Varying the seasonal drawdown limit from 10 to 30 ft resulted in total withdrawal increasing from approximately 52,000 to approximately 57,000 acre-ft; varying the year-to-year drawdown limit from 2 to 8 ft results in total withdrawal varying from approximately 53,000 to approximately 57,000 acre-ft. Increasing the minimum amount of withdrawal in the fourth quarter also affected the optimization results. Varying the fourth-quarter water demand from about 23,000 to about 41,000 acre-ft resulted in total withdrawal decreasing from about 56,000 to about 49,000 acre-ft; when the seasonal-demand constraint was increased to about 45,000 acre-ft, the optimization model was infeasible, indicating that volume cannot be pumped without violating one of the constraints. Finally, the optimization model was modified to test the impact of including groundwater-discharge constraints for the upper Lost River. The limit of these constraints was adjusted from 6 to 50 percent of baseline groundwater discharge, resulting in total withdrawal increasing from about 56,000 to about 60,000 acre-ft. For all constraint types tested in the sensitivity analyses, the optimal solution varied in a nonlinear manner over the range of constraint bounds tested.

The simulation-optimization model and its applications for the upper Klamath Basin provide an improved understanding of how the groundwater and surface-water system responds to sustained groundwater pumping within the Bureau of Reclamation's Klamath Project. Optimization model results indicate that additional pumping within the project area could be managed to minimize impact on the groundwater discharge that supports wildlife habitat in the upper Klamath Basin. For all scenarios tested, the reduction in groundwater discharge resulting from increased pumping was less than 0.2 percent, which is well within the 6-percent limit defined in the Klamath Basin Restoration Agreement. The results of the different applications of the model demonstrate the importance of identifying constraint limits in order to better define the amount and distribution of groundwater withdrawal that is sustainable. The analyses in the demonstration case presented in this report are limited by the assumption of steady average climate conditions. It is critical to note that optimal groundwater pumping volumes and patterns will change when historic hydrologic variability and the effects of nearby off-project supplemental irrigation pumping are included in the simulation-optimization model. Because these factors are not included, the pumping volumes presented may overestimate true optimal values and are not intended to be used for management decisions.

Next steps in the application of groundwater modeling in the upper Klamath Basin could include refinement of the groundwater flow model to better simulate processes and conditions in key areas of management concern. Actual application of groundwater management models will require refinement of groundwater management objectives and constraints in consultation with water users and resource management agencies, and incorporation of realistic climate variability and background supplemental pumping.

Acknowledgments

The authors acknowledge the residents and water users of the upper Klamath Basin for allowing access to their property and wells for data collection, and for sharing their observations and insights. We are grateful to Dave Sherrod (U.S. Geological Survey) and Michael Cummings (Portland State University) for critical discussions of the geology of the region, and Lenny Orzol (U.S. Geological Survey) for developing tools to efficiently process the large datasets and for his help with data visualization. Valuable hydrologic data and insights were provided by Jonathan La Marche (Oregon Water Resources Department), Tim Mayer (U.S. Fish and Wildlife Service), and Jason Cameron (Bureau of Reclamation). This work was funded by the USGS Cooperative Water Program, USGS National Research Program, the Bureau of Reclamation, and the Oregon Water Resources Department.

References Cited

Adam, D.P., Bradbury, J.P., Rieck, H.J., and Sarna-Wojcicki, A.M., 1990, Environmental changes in the Tule Lake basin, Siskiyou and Modoc Counties, California, from 3 to 2 million years before present: U.S. Geological Survey Bulletin 1933, 13 p.

Ahlfeld, D.P., and Baro-Montes, G., 2008, Solving unconfined groundwater flow management problems with successive linear programming: Journal of Water Resources Planning and Management, v. 134, no. 5, p. 404–412.

Ahlfeld, D.P., and Mulligan, A.E., 2000, Optimal management of flow in groundwater systems: San Diego, Academic Press, 185 p.

Anderson, M.P., and Woessner, W.W., 1992, Applied groundwater modeling—Simulation of flow and advective transport: Academic Press, San Diego, 381 p.

Barlow, P.M., and Dickerman, D.C., 2001, Numerical-simulation and conjunctive-management models of the Hunt-Annaquatucket-Pettaquamscutt stream-aquifer system, Rhode Island: U.S. Geological Survey Professional Paper 1636, 88 p.

Benson, S.M., Sammel, E.A., Solbau, R.D., and Lai, C.H., 1984a, Interpretation of aquifer test data, in Sammel, E.A., ed., Analysis and interpretation of data obtained in tests of the geothermal aquifer at Klamath Falls, Oregon: U.S. Geological Survey Water-Resources Investigations Report 84–4216, p. 5.1–5.55.

Benson, S.M., Janik, C.J., Long, D.C., Solbau, R.D., Lienau, P.J., Culver, G.G., Sammel, E.A., Swanson, S.R., Hart, D.N., Yee, Andrew, White, A.F., Stallard, M.L., Brown, S.P., Wheeler, M.C., Winnett, T.L., Fong, Grace, and Eakin, G.B., 1984b, Data from pumping and injection tests and chemical sampling in the geothermal aquifer at Klamath Falls, Oregon: U.S. Geological Survey Open-File Report 84–146, 101 p.

Braunsten, S.B., 2009, Subsurface structure of the Lower Klamath Lake and Tule Lake basins, California, investigated using gravity anomalies: Portland State University, Oregon, M.S. Thesis, 112 p.

Burt, Charles, and Freeman, Beau, 2003, Klamath Basin investigation—Hydrologic assessment of the Upper Klamath Basin, Issues and opportunities, Draft report: San Luis Obispo, California Polytechnic State University, Irrigation Training and Research Center (ITRAC), Prepared for the U.S. Bureau of Reclamation, May 2003, variously paged.

California Department of Water Resources, 1963, Northeastern counties ground-water investigation, Volume I–Text: State of California Department of Water Resources Bulletin 98, 246 p.

Canadell, J., Jackson, R.B., Ehleringer, J.R., Mooney, H.A., Sala, O.E., and Schulze, E. -D., 1996, Maximum rooting depth of vegetation types at the global scale: Oecologia, v. 108, p. 583–595.

Carlson, H.L., and Todd, Rodney, 2003, Effects of the 2001 water allocation decisions on the agricultural landscape and crop production in the Klamath Reclamation Project, *in* Braunworth, W.S., Jr., Welch, Teresa, and Hathaway, Ron, eds., Water allocation in the Klamath Reclamation Project—An assessment of natural resource, economic, social, and institutional issues with a focus on the upper Klamath Basin: Oregon State University Extension Service Special Report 1037, p. 163–167.

Danskin, W.R., and Freckleton, J.R., 1989, Ground-water-flow modeling and optimization techniques applied to high-ground-water problems in San Bernardino, California: U.S. Geological Survey Open-File Report 89–75, 14 p.

Danskin, W.R., and Gorelick, S.M., 1985, A policy evaluation tool—Management of a multiaquifer system using controlled stream recharge: Water Resources Research, v. 21, no. 11, p. 1731–1747.

DeSimone, L.A., Walter, D.A., Eggleston, J.R., and Nimiroski, M.T., 2002, Simulation of ground-water flow and evaluation of water-management alternatives in the upper Charles River Basin, Eastern Massachusetts: U.S. Geological Survey Water-Resources Investigation Report 02–4234, 94 p.

Doherty, J., 2010, PEST, Model-independent parameter estimation—User manual (5th ed., with slight additions): Brisbane, Australia, Watermark Numerical Computing.

Driscoll, F.G., 1986, Groundwater and wells: Johnson Division, St. Paul, Minn., 1089 p.

Fetter, C.W., 1980, Applied hydrogeology: Columbus, Ohio, Charles E. Merrill Publishing Co., 488 p.

Freeze, R.A., and Cherry, J.A., 1979, Groundwater: Prentice-Hall, Englewood Cliffs, N.J., 604 p.

Gannett, M.W., and Lite, K.E., Jr., 2004, Simulation of regional ground-water flow in the upper Deschutes Basin, Oregon: U.S. Geological Survey Water-Resources Investigations Report 03–4195, 84 p.

Gannett, M.W., Lite, K.E., Jr., La Marche, J.L., Fisher, B.J., and Polette, D.J., 2007, Ground-water hydrology of the upper Klamath Basin, Oregon and California: U.S. Geological Survey Scientific Investigations Report 2007–5050, 84 p.

Gay, T.E., Jr., and Aune, Q.A., 1958, Geologic map of California, Olaf P. Jenkins edition, Alturas sheet: California Division of Mines and Geology, 2 sheets, scale 1:250,000.

Gill, P.E., Murray, W., and Wright, M. H., 1981, Practical optimization: New York, Academic Press, 401 p.

Gorelick, S.M., 1983, A review of distributed parameter groundwater management modeling methods: Water Resources Research, v. 19, no. 2, p. 305–319.

Gorelick, S.M., Freeze, R.A., Donohue, D., and Keely, J.F., 1993, Groundwater contamination—Optimal capture and containment: Boca Raton, Fla., Lewis Publishers, 385 p.

Granato, G.E., and Barlow, P.M., 2004, Effects of alternative instream-flow criteria and water-supply demands on ground-water development options in the Big River Area, Rhode Island: U.S. Geological Survey Scientific Investigations Report 2004–5301, 110 p.

Grondin, G.H., 2004, Ground water in the eastern Lost River sub-basin, Langell, Yonna, Swan Lake, and Poe Valleys of Southeastern Klamath County, Oregon: Oregon Water Resources Department Ground Water Report No. 41, 171 p., 5 plates.

Hammond, P.E., 1983, Volcanic formations along the Klamath River near Copco Lake, Siskiyou County: California Geology, v. 36. no. 5, p. 99–109.

Harbaugh, A.W., Banta, E.R., Hill, M.C., and McDonald, M.G., 2000, MODFLOW-2000, the U.S. Geological Survey modular ground-water model—User guide to modularization concepts and the Ground-Water Flow Process: U.S. Geological Survey Open-File Report 00–92, 121 p.

Hill, M.C., 1990, Preconditioned conjugate-gradient 2 (PCG2), a computer program for solving ground-water flow equations: U.S. Geological Survey Water-Resources Investigations Report 90–4048, 43 p.

Hill, M.C., 1992, A computer program (MODFLOWP) for estimating parameters of a transient, three-dimensional, ground-water flow model using nonlinear regression: U.S. Geological Survey Open-File Report 91–484, 358 p.

Hill, M.C., 1998, Methods and guidelines for effective model calibration: U.S. Geological Survey Water-Resources Investigations Report 98–4005, 90 p.

Hill, M.C., Banta, E.R., Harbaugh, A.W., and Anderman, E.R., 2000, MODFLOW-2000, the U.S. Geological Survey modular ground-water model—User guide to the Observation, Sensitivity, and Parameter-Estimation Processes and three post-processing programs: U.S. Geological Survey Open-File Report 00–184, 210 p.

Hill, M.C., and Tiedeman, C.R., 2007, Effective groundwater model calibration, with analysis of sensitivities, predictions, and uncertainty: New York, Wiley, 455 p.

Hillier, F.S., and Lieberman, G.J., 1980, Introduction to operations research: San Francisco, Holden-Day Inc., 829 p.

Hubbard, L.H., 1970, Water budget of Upper Klamath Lake, southwestern Oregon: U.S. Geological Survey Hydrologic Investigations Atlas HA–351, 1 sheet, scale 1:250,000.

Ingebritsen, S.E., Sherrod, D.R., and Mariner, R.H., 1992, Rates and patterns of groundwater flow in the Cascade Range volcanic arc, and the effect on subsurface temperatures: Journal of Geophysical Research, v. 97, no. B4, p. 4599–4627.

Jenks, M.D., 2007, Geologic compilation map of part of the Upper Klamath Basin, Klamath County, Oregon: Oregon Department of Geology and Mineral Industries Open-File Report O-07-05, 7 p., scale 1:100,000.

Klamath Basin Restoration Agreement (KBRA), 2010, Klamath Basin restoration agreement for the sustainability of public and trust resources and affected communities— February 18, 2010: Klamath Basin Restoration Agreement, 378 p., accessed January 17, 2012, at http:// klamathrestoration.gov/.

La Rue, E.C., 1922, Klamath River and its utilization: Report to the Office of the State Engineer, Salem, Oregon, 204 p.

Leavesley, G.H., Lichty, R.W., Troutman, B.M., and Saindon, L.G., 1983, Precipitation-runoff modeling system— User's manual: U.S. Geological Survey Water-Resources Investigations Report 83–4238, 48 p.

LINDO Systems, Inc., 2005, LINDO API: Chicago, LINDO Systems, Inc., 459 p.

Loy, W.G., Allan, Stuart, Buckley, Aileen, and Meacham, Jim, 2001, Atlas of Oregon: Eugene, University of Oregon Press, 320 p.

MacLeod, N.L., and Sherrod, D.R., 1992, Reconnaissance geologic map of the west half of the Crescent 1° by 2° quadrangle, central Oregon: U.S. Geological Survey Miscellaneous Investigations Series Map I–2215, 1 sheet.

Manga, Michael, 1996, Hydrology of spring-dominated streams in the Oregon Cascades: Water Resources Research, v. 32, no. 8, p. 2435–2439.

Manga, Michael, 1997, A model for discharge in spring-dominated streams and implications for the transmissivity and recharge of Quaternary volcanics in the Oregon Cascades: Water Resources Research, v. 33, no. 8, p. 1813–1822.

Markstrom, S.L., Niswonger, R.G., Regan, R.S., Prudic, D.E., and Barlow, P.M., 2008, GSFLOW—Coupled ground-water and surface-water flow model based on the integration of the Precipitation-Runoff Modeling System (PRMS) and the Modular Ground-Water Flow Model (MODFLOW-2005): U.S. Geological Survey Techniques and Methods 6–D1, 240 p.

McDonald, M.G., and Harbaugh, A.W., 1988, A modular three-dimensional finite-difference ground-water flow model: U.S. Geological Survey Techniques of Water-Resources Investigations, book 6, chap. A1, 586 p.

McFarland, W., Gannett, M.W., Risley, J., Lynch, D., Miller, J., McCarthy, K., Snyder, D., and Morgan, D., 2005, Assessment of the Klamath project pilot water bank—A review from a hydrologic perspective: Prepared by the U.S. Geological Survey Oregon Water Science Center, Portland, Oreg.

Mehl, S.E., and Hill, M.C., 2001, MODFLOW-2000, the U.S. Geological Survey modular ground-water model—User guide to the LINK-AMG (LMG) Package for solving matrix equations using an algebraic multigrid solver: U.S. Geological Survey Open-File Report 01–177, 33 p.

Mertzman, S.A., Jr., 2000, K-Ar results from the southern Oregon-northern California Cascade Range: Oregon Geology, v. 62, no. 4, p. 99–122.

National Marine Fisheries Service, 2002, Biological opinion, Klamath project operations: National Marine Fisheries Service, Southwest Region, accessed January 5, 2007, at http://swr.nmfs.noaa.gov/psd/klamath/ KpopBO2002finalMay31.pdf.

National Marine Fisheries Service, 2010, Biological opinion, Operation of the Klamath Project between 2010 and 2018: National Marine Fisheries Service, Southwest Region, File Number: 151422SWR2008AR00148, 226 p., accessed August 29, 2011, at http://www.usbr.gov/mp/kbao/ operations_planning.html.

Natural Resources Conservation Service, 2004, Summary of the upper Klamath Basin rapid subbasin assessment of private lands: U.S. Department of Agriculture, Natural Resources Conservation Service, Oregon and California Planning Teams, 13 individually paginated chapters.

Northwest Geophysical Associates, Inc., 2002, Gravity survey groundwater investigation Lower Klamath Lake Basin Klamath County, Oregon: Corvallis, Oreg., Northwest Geophysical Associates, Inc., prepared for Klamath Drainage District, May 2002, 13 p. plus 11 sheets.

Oregon Administrative Rules, 2011, Oregon Water Resources Department, Administrative Rules, Chapter 690, Division 8, Statutory Ground Water Terms, accessed August 30, 2011, at http://www.wrd.state.or.us/OWRD/LAW/oar.shtml.

Oregon State University PRISM Group, 2006, PRISM Group data website: accessed September 2006 at http://www.ocs.oregonstate.edu/prism.

Orr, E.L., Orr, W.N., and Baldwin, E.M., 1992, Geology of Oregon: Kendall/Hunt Publishing Co., 254 p.

Prudic, D.E., 1989, Documentation of a computer program to simulate stream-aquifer relations using a modular, finite-difference, ground-water flow model: U.S. Geological Survey Open-File Report 88–729, 113 p.

Reichard, E.G., Land, M., Crawford, S.M., Johnson, T., Everett, R.R., Kulshan, T.V., Ponti, D.J., Halford, K.J., Johnson, T.A., Paybins, K.S., and Nishikawa, T., 2003, Geohydrology, geochemistry, and ground-water simulation-optimization of the central and west coast basins, Los Angeles County, California: U.S. Geological Survey Water-Resources Investigation Report 03–4065, 184 p.

Sammel, E.A., and Peterson, D.L., 1976, Hydrologic reconnaissance of the geothermal area near Klamath Falls, Oregon, with a section on Preliminary interpretation of geophysical data: U.S. Geological Survey Water-Resources Investigations Report 76–127, 137 p.

Shah, Nirjhar, Nachabe, Mahmood, and Ross, Mark, 2007, Extinction depth and evapotranspiration from ground water under selected land covers: Ground Water, v. 45, no. 3, p. 329–338.

Sherrod, D.R., 1991, Geologic map of a part of the Cascade Range between latitudes 43°–44°, central Oregon: U.S. Geological Survey Map I–1891, 14 p., scale 1:125,000, 1 sheet.

Sherrod, D.R., and Pickthorn, L.G., 1992, Geologic map of the west half of the Klamath Falls 1° by 2° quadrangle, south-central Oregon: U.S. Geological Survey Miscellaneous Investigations Series, Map I–2182, 1 sheet, scale 1:250,000.

Sherrod, D.R., and Smith, J.G., 2000, Geologic map of upper Eocene to Holocene volcanic and related rocks of the Cascade Range Oregon: U.S. Geological Survey Miscellaneous Investigations Map I–2569, scale 1:500,000.

Smith, J.G., Page, N.J., Johnson, M.G., Moring, B.C., and Gray, Floyd, 1982, Preliminary geologic map of the Medford 1° by 2° quadrangle, Oregon and California: U.S. Geological Survey Open-File Report 82–955, 1 sheet, scale 1:250,000.

Snyder, D.T., and Morace, J.L., 1997, Nitrogen and phosphorous loading from drained wetlands adjacent to Upper Klamath and Agency Lakes, Oregon: U.S. Geological Survey Water-Resources Investigations Report 97–4059, 67 p.

Theis, C.V., 1940, The source of water derived from wells—Essential factors controlling the response of an aquifer to development: Civil Engineering, v. 10, p. 277–280.

U.S. Fish and Wildlife Service, 2008, Biological/Conference opinion regarding the effects of the U.S. Bureau of Reclamation's proposed operation plan for the Klamath Project and its effects on the endangered Lost River and shortnose suckers.

Vance, J.A., 1984, The lower Western Cascades Volcanic Group in northern California, in Nilsen, T.H., ed., Geology of the Upper Cretaceous Hornbrook Formation, Oregon and California: Society of Economic Paleontologists and Mineralogists, Pacific Section, Field Trip Guidebook v. 42, p. 195–196.

Veen, C.A., 1981, Gravity anomalies and their structural implications for southern Oregon Cascade Mountains and adjoining Basin and Range province: Corvallis, Oregon, Oregon State University, M.S. thesis, 86 p.

Wagner, B.J., 1995, Recent advances in simulation-optimization groundwater management modeling: Reviews of Geophysics, Supplement 33, p. 1021–1028.

Wagner, D.L., and Saucedo, G.J., compilers, 1987, Geologic map of the Weed Quadrangle: California Division of Mines and Geology Regional Geologic Map Series Map No. 4A (Geology), 15 p., 4 sheets, scale 1:250,000.

Walker, G.W., 1963, Reconnaissance geologic map of the eastern half of the Klamath Falls (AMS) quadrangle, Lake and Klamath counties, Oregon: U.S. Geological Survey Mineral Investigations Field Studies Map MF–260, 1 sheet scale 1:250,000.

Western Regional Climate Center, 2006, Historical climate data: accessed September 2006 at http://www.wrcc.dri.edu.

Yeh, W. W-G., 1992, Systems analysis in ground-water planning and management: Journal of Water Resources Planning and Management, v. 118, no. 3, p. 224–237.

www.ingramcontent.com/pod-product-compliance
Lightning Source LLC
Chambersburg PA
CBHW081546170526
45166CB00009B/2603